T0186807

# HUMANISTIC ASPECTS OF TECHNICAL COMMUNICATION

Edited by
Paul M. Dombrowski
*Ohio University*

Baywood's Technical Communications Series
Series Editor: JAY R. GOULD

Baywood Publishing Company, Inc.
AMITYVILLE, NEW YORK

Library of Congress Catalog Card Number: 94-14937

ISBN: 0-89503-159-0 (Cloth)
ISBN: 0-89503-160-4 (Paper)

**Library of Congress Cataloging-in-Publication Data**

Humanistic aspects of technical communication / edited by P. M.
Dombrowski.
     p.  cm. - - (Baywood's technical communication series)
     Includes bibliographical references and index.
     ISBN 0-89503-159-0 (cloth). - - ISBN 0-89503-160-4 (paper)
     1. Communication of technical information- -Social aspects.
I. Dombrowski, P. M. (Paul M.) II. Series: Baywood's technical
communications series (Unnumbered)
T10.5.H86 1994
601.4- -dc20                                 94-14937
                                                    CIP

# Preface

This book is a collection of essays on the humanistic aspects of technical communication with extensive commentary. These aspects deal with the human side of technical communication rather than the technical side of information and objects. They also deal with the social context of the origin, definition, and use of information. These aspects form the chapter headings: rhetoric of science, social constructionism, feminism and gender issues, and ethics.

The first chapter deals with humanism in general as it relates to technical communication. It is followed by chapters on each of the four areas mentioned above. Each of these chapters includes an overview, critical comments, and a summary of two reprinted essays, then the essays.

The selected essays do not necessarily represent the entire current state of their areas. They instead were chosen primarily to present important statements, to suggest important issues, or to represent the general tenor of activity in their areas. My overview and critical comments are likewise not definitive. They are meant primarily as informative, provocative, and useful bases for understanding and critically examining each area.

This book has two audiences and purposes: one, graduate teaching assistants and their instructors, the other, a general readership of scholars and professionals.

The teaching audience and purpose stem from the context of origin of this work, a graduate seminar in teaching technical writing at Ohio University. This first audience has a strong literary background. The purpose for them is to introduce technical communication from a readily accessible avenue, through the same philosophy of humanism and literary communication with which they are already familiar and closely allied. This book differs then from the many traditional technical communication textbooks which assume a primarily scientific or technical background. For this audience and purpose, this collection can serve as an adjunct to conventional technical communication textbooks.

The general readership of scholars and professionals is already familiar with technical communication. The purpose for this second audience is to provide a convenient collection of significant essays on recent humanistic developments in technical communication. These developments from philosophy, rhetoric, sociology, feminism, and ethics might be only sketchily known by this audience. Many readers will be stimulated by the humanistic perspective, which insists that the root subject of technical communication is always humankind.

Thus these developments of a humanistic nature will interest both the academic and non-academic communities. Though they have great bearing on communication within fields of specialized knowledge such as engineering, they also have substantial bearing on the interface between the specialized worlds of technology and science and the broader world of general, non-specialized society. Technical communication from this humanistic perspective is seen in its unique nature as the bridge between the sciences and the humanities. This bridge spans the separation decried by C. P. Snow, the scientist and literary critic, in his famous *Two Cultures* essay. No other academic or professional field is so intrinsically involved in both cultures, and no other field holds as great potential for integrating the two.

# Acknowledgments

The following have graciously granted permission to reprint articles used in this book.

R. Allen Harris, Assent, Dissent, and Rhetoric in Science, *Rhetoric Society Quarterly, 20*, pp. 13-37, 1990. Copyright 1990 by the Rhetoric Society of America.

A. G. Gross, Discourse on Method: The Rhetorical Analysis of Scientific Texts, *Pre/Text, 9*:3-4, pp. 169-185, 1988. Copyright 1988 by Victor J. Vitanza, editor.

P. M. Dombrowski, Challenger and the Social Contingency of Meaning: Two Lessons for the Technical Communication Classroom, *Technical Communication Quarterly, 1*:3, pp. 73-86, 1992. Copyright 1992 by the Association of Teachers of Technical Writing.

C. R. Miller, Some Perspectives on Rhetoric, Science, and History, *Rhetorica, VII*:1, pp. 101-114, 1989. Copyright 1989 by International Society for the History of Rhetoric.

M. M. Lay, Feminist Theory and the Redefinition of Technical Communication, *Journal of Business and Technical Communication, 5*:4, pp. 348-370, 1991. Copyright 1991 by Sage Publications, Inc.

J. Allen, Gender Issues in Technical Communication Studies: An Overview of the Implications for the Profession, Research, and Theory, *Journal of Business and Technical Communication, 5*:4, pp. 371-392, 1991. Copyright 1991 by Sage Publications, Inc.

D. Sullivan, Political-Ethical Implications of Defining Technical Communication as a Practice, *Journal of Advanced Composition, 10*:2, pp. 375-386, 1990. Copyright 1990 by the Association of Teachers of Advanced Composition.

# Table of Contents

Preface . . . . . . . . . . . . . . . . . . . . . . . . . . . . . . . . . . .   iii

Acknowledgments . . . . . . . . . . . . . . . . . . . . . . . . . . . .   v

CHAPTER 1   HUMANISM AND TECHNICAL COMMUNICATION .   1

CHAPTER 2   RHETORIC OF SCIENCE . . . . . . . . . . . . . . . .   15

    Assent, Dissent, and Rhetoric in Science
      *R. Allen Harris* . . . . . . . . . . . . . . . . . . . . . . . . . . .   33

    Discourse on Method: The Rhetorical Analysis of Scientific Texts
      *Alan G. Gross* . . . . . . . . . . . . . . . . . . . . . . . . . . .   63

CHAPTER 3   SOCIAL CONSTRUCTIONISM . . . . . . . . . . . . . .   81

    Challenger and the Social Contingency of Meaning:
    Two Lessons for the Technical Communication Classroom
      *Paul M. Dombrowski* . . . . . . . . . . . . . . . . . . . . . .   97

    Some Perspectives on Rhetoric, Science, and History
      *Carolyn R. Miller* . . . . . . . . . . . . . . . . . . . . . . . .   111

CHAPTER 4   FEMINIST CRITIQUES OF SCIENCE AND
              GENDER ISSUES . . . . . . . . . . . . . . . . . . . . .   125

    Feminist Theory and the Redefinition of Technical Communication
      *Mary M. Lay* . . . . . . . . . . . . . . . . . . . . . . . . . . .   141

Gender Issues in Technical Communication Studies: An Overview
of the Implications for the Profession, Research, and Pedagogy
*Jo Allen* . . . . . . . . . . . . . . . . . . . . . . . . . . . . . . . 161

CHAPTER 5   ETHICS . . . . . . . . . . . . . . . . . . . . . . . . . . 181

A Basic Unit on Ethics for Technical Communicators
*Mike Markel* . . . . . . . . . . . . . . . . . . . . . . . . . . . 199

Political-Ethical Implications of Defining Technical
Communication as a Practice
*Dale L. Sullivan* . . . . . . . . . . . . . . . . . . . . . . . . . 223

Contributors . . . . . . . . . . . . . . . . . . . . . . . . . . . . . . . 235

Index . . . . . . . . . . . . . . . . . . . . . . . . . . . . . . . . . . . 237

# CHAPTER 1

# Humanism and Technical Communication

Technical communication seems to have two aspects, the technical and the humanistic. The technical aspect is concerned with specialized knowledge, conventional forms, and traditional products. It is the more obvious of the two and is specific to technical and scientific communication.[1]

The humanistic aspect (to be defined shortly), brought to prominence by C. Miller's landmark essay, *A Humanistic Rationale for Technical Writing,* is the less obvious and is not specific to technical communication, though it is no less operative [1]. It is vitally important because it deals with the humanity of the sender and receiver of the communication model and with the wide social impact of the communication. It is concerned, for example, with how people persuade each other and with how people decide the course of technological developments. It also deals with our common human nature as beings who continually re-define our worlds through language. The increasing development of this humanistic aspect yields a fuller, more balanced understanding of technical communication which includes the manifold social contingencies and ramifications of technical communication.

The thrust of recent developments of this humanistic aspect discussed in this book is to challenge the long-standing dualities of the sciences versus the humanities, facts versus opinions, and objectivity versus subjectivity, which are seen as no longer tenable absolutely. Instead, these dualities are seen conditionally as social constructions serving particular goals and purposes. Examples such as the charring of

---

[1] Throughout this volume, I will usually not distinguish between science and technology and will use "technical communication" as including both technical and scientific communication because they both deal with specialized knowledge.

the O-rings of the Challenger space shuttle show that the meaning of technical information is socially constructed and shaped by opinion, not just by "the facts themselves."

These humanistic developments also provide a framework for understanding how communication about science and technology occurs in a context of social responsibilities.[2] Thus we are faced with such questions as whether treating another person as an "object" of scientific study does not unfairly privilege the scientist observer over the relatively disempowered person observed.

The essays presented in the following chapters were chosen as important statements indicative of the tenor and the direction of developments regarding technical communication in each area: rhetoric of science, social constructionism, feminism and gender issues, and ethics. Each chapter begins with an overview of the topic accompanied by extensive critical analysis. These initial comments are meant to be informative, provocative, and useful bases for understanding and critically examining that particular topic. When they diverge from prevailing opinion, it is because I intend not only to describe current developments but also to offer constructive criticism and suggest future developments.

In this chapter, I will first discuss how these recent developments are humanistic. Then I will discuss two conceptual frameworks for understanding the relation between the humanism and technical communication, together with the view of language associated with each. This distinction indicates that technical communication is less a make-shift bridge between disparate enterprises than the intellectual common ground already shared by them.

The distinction between these frameworks is crucial for grasping the humanistic developments I will be discussing later. It is important to understand fully these frameworks and the ramifications of their differences, though initially the distinction might seem too subtle or abstract. These different frameworks amount to different world-views (especially the world of technical communication), not that the world itself changes but that how we see the world and what we understand as the world changes.

## HUMANISM IN CLASSICAL AND
## MODERN SENSES

The four areas of recent development (rhetoric of science, social constructionism, feminism and gender issues, and ethics) are closely interrelated as different

---

[2] P. M. Rubens in Technical and Scientific Writing and the Humanities reviews the history of the relation between the humanities (especially ethics and rhetoric) and scientific and technical writing, and explains the critical centrality of one's view of language in shaping such writing [2]. Rubens also provides an extensive review of the literature on this topic through 1984.

aspects of a single general frame of mind: humanism. "Humanism" is apt in two senses, classic and strict, and modern and loose. [3]

## Classical and Renaissance Classicist Thought

In the strict sense of "humanism," these four areas are recent expressions of the same frame of mind represented in the *studia humanitatis* of Italian humanism. These humanists of the fourteenth and fifteenth centuries were the initiators of the Renaissance. They were students and teachers of the rhetoric, history, moral philosophy and other areas of ancient Greek and Roman thought, keenly interested in the resurrection and expansion of these classical ideas. Among these ideas are the constitutive power of language, the social responsibilities of citizens, and the advancement of civilization through free, critical discourse.

Their purpose was to apply these ideas in order to elevate and improve their culture. As G. A. Kennedy, the historian of the rhetoric, explains, these Renaissance humanists turned to classical ideas in order to revitalize their stagnant culture [4]. They worked to recover classical rhetoric, for example, because it was a noble, creative activity that acted as the well-spring of civilization.

The four contemporary areas of this volume reflect not only the same general interests of the *studia humanitatis* but also the act of resurrection itself, the cultivation not of something new but a re-birth of what had already been born but had become practically dead; thus the re-nasence or Renaissance.

Well before the Italian humanists of course, classical Greek thinkers originated the liberal arts and humanism in another strict sense. B. Kimball, a recent proponent of the re-unification of rhetoric and science, explains that what we know today as the "arts and sciences" began in ancient Greece as the seven liberal arts which constituted the enlightened pursuits of the free citizens of Greece [5].

The pursuit of these liberal arts was thought to be freeing in itself, elevating the practitioner above baser impulses while cultivating a noble, civic mentality. These were the three arts of the *trivium* (rhetoric, grammar, and logic) and the four arts of the *quadrivium* (astronomy, arithmetic, geometry, and music). [4] Indeed, many historians consider the statement of the sophist Protagoras to be the fundamental

---

[3] J. S. Nelson and A. Megill's *Rhetoric of Inquiry: Projects and Prospects* contains an excellent synoptic history of the relations between rhetoric and philosophy generally (especially the "absolutized dichotomy between truth and opinion"), as well as among rhetoric, the humanities, and the social sciences in particular [3, p. 21].

[4] These seven arts do not of themselves constitute either humanism or the ideal of Greek education. For the Greeks, education aimed at cultivation of a universal cultural ideal which could not be reduced to these seven arts or even to any collection of arts (*techne*). W. Jaeger points out in his authoritative history of the Greeks that indeed Protagoras clearly distinguished between technical knowledge and the universal, humanistic culture named *paideia* [6].

principle of humanism: "Man is the measure of all things." Protagoras would feel right at home among contemporary social constructionists.

Thus from the earliest history of Western higher education, "humanistic" meant the liberal arts which embraced *both* the sciences (astronomy, arithmetic, geometry, and logic) and what we call today the humanities (rhetoric, grammar, and music and its allied arts). To be a humanist historically meant to be interested simultaneously in the specialized knowledge of the sciences and in the social context in which specialized knowledge was put into practice through rhetoric. [5]

## Modern Thought

In a modern, looser sense, too, "humanistic" is apt for these developments. Broadly speaking, humanism is the emphasis of the human over the non-human. It involves studies turned more toward humankind itself than toward the physical, non-human world, for example, toward literature or ethics. This humanistic emphasis is reflected in technical communication studies as interest in persuasion, in psychological and sociological constructs, in human technological practices, and in gender-fairness—all of these emphasized over the objective, material, non-human things that are often taken as the basic subject of technical communication. Thus, from the perspective of humanism, the root subject of technical communication is always humankind.

The humanistic aspects of technical communication are not, let me reiterate, new additions to our field but only the growing recognition of previously unacknowledged aspects of what has always been there. The rhetoric of science, for example, reveals the paradoxical pathos of dispassion in science. As A. G. Gross explains, this apparent unemotionalness is only a disguise for high emotionality: "[T]he disciplined denial of emotion in science is only a tribute to our passionate investment in its methods and goals" [8, p. 179].

Feminist critiques of science are another example of making apparent what had been inapparent. The traditional view of science is that it is value-free, as though it somehow transcends moral values or is at least ethically indifferent. Some feminists (see Chapter 4 on feminist critiques of science), however, point out that some scientists view human behavior as principally determined by sex, a biological feature that can readily be studied scientifically. These feminists point out that

---

[5] S. M. Halloran's *Rhetoric in the American College Curriculum: The Decline of Public Discourse* links classical rhetoric to more recent times [7]. He explains that the earliest American higher education was informed by the classical view of rhetoric by which rhetoric was understood as a master art and a cultural ideal by which specialized knowledge was made relevant and applied to issues of public concern. Only since about the nineteenth-century has the valuation of rhetoric declined concomitantly with the elevation of the specialized knowledge of the sciences. The result of this shift in valuation, Halloran explains, has been the vigorous pursuit of specialized, arcane knowledge for its own sake and the decline of public discourse by which civic needs are given priority and important issues are debated in a public forum.

an exclusively scientific perspective on questions implicitly denies the vitally important effects of culture and personal assent, effects which are highly amorphous and difficult to treat "scientifically." The result has been a neglect (and often an implicit devaluing) of major factors in human behavior by those people, scientists, to whom the public turns for answers.

The thrust of recent humanistic developments, then, is to highlight otherwise obscured elements in technical communication. This highlighting shows the historical duality between the sciences and the humanities to be largely untenable. Gross points out that over the past two decades, the intellectual world has seen a "blurring of genres," an intermixing of intellectual disciplines previously seen as disparate. More specifically, Gross explains that his own investigations of the rhetoricity of science are intended to mend the dualistic rupture (originating in Plato's attacks on sophism and fostered by Cartesianism) between the sciences and the humanities by affirming "the permanent bond that must exist between science and human needs" [8, p. 183].

## CONCEPTUAL FRAMEWORKS

In this section, I will discuss two conceptual frameworks for understanding the relation between technical communication and humanism: traditional dualism and contemporary holism. Each framework has a particular view of language associated with it. The traditional dualistic framework entails the transparency view of language, while the contemporary holistic framework entails the rhetorical view of language. Though many traditional technical communicators conceive of our field from the first framework, the humanistic developments I will be reviewing argue for a re-conceptualization of our field along the lines of the second framework. Depending on the framework, one sees technical communication as straddling polar opposites, or as the common ground between related enterprises that differ more in emphasis than in kind.

### Traditional Dualism

In this section, I will describe the traditional dualistic framework, outline its history, and characterize the view of language which it entails.

#### Description

From the traditional perspective, technical communication has a unique nature with respect to the humanities. Though technical communication sometimes occurs between specialists within a specialized community, it also often occurs between specialists and nonspecialists, say, between technical experts and the nontechnical, nonscientific public. In this form, technical communication is the principal field bridging two realms traditionally thought of as separate and different, so radically different, in fact, that they are often seen as having little to do with each other. The usual names for these two realms are "the sciences" and "the humanities" and for the partitioning into separate entities is "dualism."

Though technical communication is not the only field in which the sciences and the humanities co-mingle, it is unique in pointedly making accessible to non-specialists the specialized knowledge of science and technology. In other areas, such as engineering, facts can be allowed to predominate over opinions (obscuring the role of negotiated opinion in establishing facts), while in ethics, say, values and belief can be allowed to predominate over facts (obscuring the role of material consequences in shaping opinion).

From this perspective, the bridge of technical communication must be grounded in both the world of science and technology and in the world of the humanities. To perform its function well, then, technical communication must be informed by humanistic studies. [6]

*History*

The history of this conceptual framework involves two major developments, ancient Greek rhetoric and modern empirical science.

In the classical period, Plato based his renowned tirades against rhetoric on the mutability of opinion. Because opinion is notoriously mutable, he held, it cannot be true in any absolute sense and so has no intrinsic merit. Merit, for Plato, lies only in what is immutable, which must therefore be a reflection of the divine precisely because of its elevation above mutable human opinion.

Plato's student, Aristotle, continued Plato's distinction between truth and opinion though not his animosity toward rhetoric. For Aristotle, it was the role of distinctly nonrhetorical activities to find the unchanging and unequivocal, the absolute and the true. These activities were the sciences, rigorous studies seeking only the demonstrably and immutably true. [7] Thus Aristotle distinguished between *doxa* (opinion) and *episteme* (true knowledge). [8]

---

[6] M. S. Samuels discusses the relation between the sciences and the humanities from the point of view of the philosophy of Giambattista Vico [9]. Vico understood the history of the prevailing authority of knowledge as the periodic alternation between the sciences and the humanities. Samuels identifies the works of Halloran, Zappen, Miller, Whitburn, Rutter, Kuhn (all discussed later) and a host of others as indicative of a swing in recent times toward humanism.

[7] Another important usage of "humanism" defines it as the attitude of rigorous, systematic study open to alteration, studies which are opposed to superstition and to the authority of received knowledge. From this perspective, science is itself an expression of humanism with roots extending back to the Pre-Socratic philosophers. See Kohanski, Kimball, or Bronowski for histories of this train of thought [10, 5, 11].

[8] Aristotle's and Plato's were not the only classical rhetorics. Isocrates, for instance, valorized rhetoric as the thrashing out of authoritative knowledge and as the very essence of our humanness, the faculty which raises us above other beings. Aristotle's view, nonetheless, has traditionally been the most prominent.

The radical distinction between truth and opinion drawn by Plato and Aristotle was drawn with a vengeance in later periods. Ramism in the Middle Ages, for instance, elevated rationalism to an obsession with logic and absolute truth while it reduced rhetoric to only empty ornamentation.

Empirical science, which arose following the Renaissance, further widened the separation between opinion and truth, rhetoric and science, into a chasm. Francis Bacon and the Royal Society created what we now call modern empirical science. These thinkers vehemently denounced rhetoric and defined the emerging empirical Science as specifically opposed and impervious to opinion [11]. S. M. Halloran points out, for instance, that they deemed Science to be majestically above the need of the empty ornamentation of rhetoric, as almost preternaturally elevated above the foibles and fallibility of humanity. Echoing Plato in insisting on the incontestably true, then, Baconian Science sought what was greater than the baser aspects of humanity, including the shifting babble of language. [9]

*View of Language*

The view of language associated with this traditional dualism is the "transparency" view of language (Miller's Humanistic Rationale offers a good characterization of this view [1]). This is the view that the ideal communication transparently and directly represents to the receiver the external, objective world of facts without distortion by language. Thus good language use is thought to work like a windowpane, providing direct, undistorted perception of objective reality by the observer. Language would thus be, paradoxically, a medium which provides unmediated perception. This view entails two assumption, both problematic from the humanistic perspectives discussed here. First, it assumes that the objective world can be known directly and with certainty. Second, it assumes that this knowledge can be separate from language and theories articulated in language. Both assumptions are questioned, we will see, by rhetoricians of science and social constructionists.

A good example of this framework at work is the technical communication textbook by B. E. Cain, a professor of chemistry. Cain's *The Basics of Technical Communicating* carries the authority of the American Chemical Society as one of its authoritative Professional Reference Books [14]. Cain states that technical communication differs from other communication in being concerned primarily with facts. It is the purpose of technical communication, he says, to present or relay already-defined, pre-existing information to various audiences and to do this

---

[9] The juxtaposition of Baconian Science and Platonic Idealism may seem jarring to some. R Rorty, the contemporary philosopher of pragmatism, explains, however, that the relation between science and idealism stems from the search for generality. This generality necessarily transcends both particularity and materiality and therefore constitutes an idealism [13, p. 43]. B. Kimball, the philosopher, echoes the parallel between modern science and idealism [5].

with as little emotion, subjectivity, or opinion as possible. Though on the face of it this understanding of the role of language seems unproblematic and obviously desirable, many contemporary thinkers, as we will see, challenge the assumptions behind this statement and offer an alternative framework based on different assumptions.

The transparency view strikes such critics as needlessly suspicious of language while denying language the capacity to constitute knowledge and meaning. This suspicion and denial has the effect of distancing the sciences and technology from the humanities and from language-centered knowledge and meaning such as is found in traditional English departments. As R. Rutter puts it, "Gradually it became the norm to assume that so-called 'hard' disciplines would supply the cake of content, while departments of English would supply the frosting of style" [15, p. 143]. For those holding the transparency view, the only contribution the English department can make is without substance or real significance.

## Contemporary Holism

In this section I will describe this framework, outline its history, and characterize the view of language which it entails. To distinguish it from traditional dualism, I will include examples of how this view of language blurs the distinction between fact and opinion.

### Description

Holism means seeing as a unitary whole what is otherwise seen as separate and opposite. A good example is the mind-body dualism implicit in traditional scientific medicine. A holistic perspective, such as psychosomatic medicine, sees the mind and body as a single interrelated whole, the conceptual separation of which is without true usefulness. A holistic view of science and technology sees fact and opinion or language and knowledge as twin aspects of a whole rather than as radically separate.

Indeed, these contemporary developments undermine the assumption of separation between the sciences and the humanities. From a humanistic perspective, there is no separateness. The holistic humanistic perspective is also innovative in questioning if not undermining the specialness of the specialized knowledge of technicians and scientists, including the claim to authoritativeness over other sorts of knowledge (such as intuition) and over opinion.

### History

Though holism is a very old idea, its recent resurgence is of primary interest to us. This resurgence is of two kinds: general intellectual developments spanning many fields and specific developments in communication.

In our century, the intellectual and spiritual impoverishment of higher education suffered with the loss of the elements of moral philosophy such as politics, ethics, and rhetoric has become increasingly recognized. For instance, C. P. Snow, himself both scientist and literary critic, in his famous *Two Cultures* lecture decried the separation between the sciences and the humanities as unwarranted and unfruitful [16]. J. Dewey, a major American philosopher and educator, also was deeply concerned about the aridity, alienation, and loss of sense of community that has resulted from an exclusive concern with science and technology and the impersonality that characterized them [17].

In recent years, the affirmation of the inseparability of the two realms has become almost a commonplace in academic circles. Even within science, R. Carson, the biologist who pioneered ecological awareness, L. Pauling, the Nobel-laureate chemist who argues against the duality of mind and body, and R. Dubos, the microbiologist and social critic, have urged the tempering of scientific development with greater concern for the human, social effects of these developments. Philosophers and sociologists of science such as Kuhn, Merton, Watzlawick, Latour, Woolgar, and Berger and Luckmann (discussed in later chapters on rhetoric of science and social constructionism) revealed the surprising rhetoricity and social contingency of scientific knowledge itself. In philosophy, M. Heidegger, perhaps the greatest European philosopher of our century, expressed grave concern about the privilege accorded science and technology, a concern seconded by later American social pragmatic philosophers such as H. Putnam and R. Rorty.

Within popular culture, J. Ellul, the French critic of the ethics of technology, has criticized the absence of restraints on the expansion of science and technology [18]. L. Winner, the American sociologist of science, echoes similar concerns about the seemingly autonomy of technology [19]. In education, B. Kimball, has critiqued the unbalanced primacy given to philosophy in the form of contemporary science, and called for the counterbalancing re-vitalization of rhetoric and civic leadership to accommodate science to humanity [4].

The communication fields have undergone parallel developments. Regarding technical communication, C. R. Miller's urges a counterbalancing of an exclusive preoccupation with the technical side of technical communication by an emphasis on the humanistic, "communitarian" side [13]. S. M. Halloran and M. Whitburn separately have voiced similar concerns, pointing out the impoverished view of language traditionally associated with technical communication as well as the potential for detachment from ethics and civic responsibility which such a view fosters [11, 20]. C. Bazerman reveals the social processes by which scientific communications are shaped [22]. K. Burke LeFevre urges the active, responsive, and responsible participation of the technical communicator in the constitution of knowledge in an interactive relationship between society and science and technology [21]. [10]

---

[10]Of a broader nature are articles by D. Sullivan (ethics and politics), D. Bradford (empowerment through humanism), R. Rutter (call for humanistic approaches), G. Parsons (general education), E. Harris (liberal arts approach), R. Rutter (role of imagination), M. S. Samuels (literary reality), and A. Manning (recreation vs. substitution of reality) [23-30].

*View of Language*

The view of language associated with the humanistic framework emphasizes the holistic interrelatedness of fact and opinion, the subjectivity of knowledge, and the rhetorical negotiation of knowledge. It asserts that mind and matter, subject and object, fact and opinion are inherently inseparable. For simplicity, I will call this the rhetorical view of language.

Perhaps we can understand the rhetorical view of language best by focusing on two aspects: the interrelatedness of fact and opinion, and the subjectivity of knowledge.

Rather than perpetuating the duality of facts *or* opinions, technical communication must recognize the conditional usefulness of both terms, "fact" *and* "opinion." This frame of mind sees both terms not as crisp-edged, exclusive entities but only as significations which always, in varying degrees, include elements of the other. We can think of this as a continuum the end points of which represent the maximal degree of facticity or opinion-ness. In this view, the communicator must continually be aware of both the facticity and the opinionness of knowledge, acknowledge the interactive interplay between the two, and take pains explicitly to discuss these aspects of his or her communication.

## Examples

Regarding the Challenger disaster, I have explored the notions of "anomalousness" and "flightworthiness" and shown how these apparent facts were highly socially contingent [31]. Whether the charring of the shuttle O-rings was to be considered an anomaly or not was itself a matter of debate. Also, the more investigators probed vigorously for additional facts, the less it was clear to them what the facts were or at least what the facts meant, and the less valid became the separation of fact from opinion. Indeed, W. B. Weimer explains that scientific facts in general are "structurally ambiguous: one and the same surface-structure entity is seen to have two (or more) deep structural representations or meanings (factual attributions)" [32, p. 6].

The rhetorical nature of scientific and technical language is also revealed through the subjectivity of science. L. Fleck, perhaps the first sociologist of science, noticed the subjectivity of "objective" science in his studies of the notion of syphilis as a disease. These studies later stimulated T. Kuhn to formulate his revolutionary sociology of knowledge [33, 34]. Fleck explained that the "facts" of the existence, definition, and transmission of syphilis constantly shift with opinion and social mores.

In recent years, P. Feyerabend, the iconoclastic philosopher of science, has revealed the powerful involvement of Galileo's psychological, cognitive assumptions in his observations of the moons of Jupiter, assumptions which gave shape to his observations [35]. Thus what ordinarily is taken as a highly objective observation was really a construction. Contrary to the usual conceptualization of the relation between observational facts and theory, Feyerabend also carefully revealed that Galileo's "facts" of observed revolution around Jupiter were conditioned by Galileo's prior theoretical/conceptual assumption that such an arrangement was possible and plausible. For Galileo, then, facts were constituted by his prior opinions, opinions which amounted to theories. Thus, theory did not follow from observations but preceded them.

These examples show that facts are both defined and constituted by opinion. They also illustrate that though opinions can be constituted from facts, in a surprising way facts can also be constituted from opinions in the form of preconceptions. Therefore the line of demarcation between facts and opinions is never absolutely clear and can never be taken for granted. As we will see in the following chapters, some critics go so far as to assert the inescapable subjectivity of all objectivity.

## CONCLUSION

The prospective result of recent developments is to redefine the field of technical communication as an organic whole affirming the fundamental humanity of both communicators and science and technology. Thus the basic thrust of this volume is to challenge the interrelated dualities of fact versus opinion and the sciences versus the humanities, dualities which seem to critics to dehumanize science while they disempower the humanities. Recognizing the very human rhetorical and social contingency of all knowledge and opinion is an important step toward a world informed by science and technology in service to humankind.

In the following two chapters we will see, for instance, that studies of the rhetorical and social workings of science and technology open them up to criticism and to deliberated alteration. Later chapters then specify some of these criticisms, for example in feminists's objections to the dominance of males in science or in ethicists's concerns that technology not be perceived as autonomous, unbridled in pursuing its own imperative.

More concretely, each of these topics is becoming an important part of technical communication textbooks. Every new textbook discusses ethics; most discuss if not emphasize social aspects such as collaborative writing; and many discuss women's concerns as well. Whole books are devoted to the rhetoric of science (e.g., A. Gross's *The Rhetoric of Science*), to social constructionism (e.g., C. Thralls and N. Roundy Blyler's (eds.) *The Social Perspective and Professional Communication*), and to feminist critiques of science (e.g., S. Harding's *The Science Question in Feminism*). For these important new developments, this book can serve as an introduction and critical review.

## REFERENCES

1. C. R. Miller, A Humanistic Rationale for Technical Writing, *College English, 40*:6, pp. 610-617, 1979.
2. P. M. Rubens, Technical and Scientific Writing and the Humanities, in *Research in Technical Communication*, M. G. Moran and D. Journet (eds.), Greenwood Press, Westport, Connecticut, 1985.
3. J. S. Nelson and A. Megill, Rhetoric of Inquiry, *Quarterly Journal of Speech, 72*, pp. 20-37, 1986.
4. G. A. Kennedy, *Classical Rhetoric and Its Christian and Secular Tradition from Ancient to Modern Times*, University of North Carolina Press, Chapel Hill, North Carolina, 1980.
5. B. Kimball, *Orators & Philosophers*, Teacher College of Columbia University, New York, 1986.
6. W. Jaeger, *Paideia: The Ideals of Greek Culture,* Vol. I, Gilbert Highet (trans.), Oxford University Press, New York, 1939.
7. S. M. Halloran, Rhetoric in the American College Curriculum: The Decline of Public Discourse, *Pre/Text, 3*:3, pp. 245-269, 1982.
8. A. G. Gross, Discourse on Method: The Rhetorical Analysis of Scientific Texts, *Pre/Text, 9*:3-4, pp. 169-185, 1988.
9. M. S. Samuels, Is Technical Communication "Literature"? Current Writing Scholarship and Vico's Cycles of Knowledge, *Journal of Business and Technical Communication, 1*, pp. 48-67, 1987.
10. A. S. Kohanski, *The Greek Mode of Thought in Western Philosophy*, Fairleigh Dickinson University Press, New York, 1984.
11. J. Bronowski, *The Ascent of Man*, Little, Brown, Boston, Massachusetts, 1974.
12. S. M. Halloran, Eloquence in a Technological Society, *Central States Speech Journal, 29*, pp. 221-227, 1978.
13. R. Rorty, *Philosophy and the Mirror of Nature*, Princeton University Press, Princeton, New Jersey, 1979.
14. B. E. Cain, *The Basics of Technical Communication, An ACS Professional Reference Book*, American Chemical Society, Washington, D.C., 1988.
15. R. Rutter, History, Rhetoric, and Humanism: Toward a More Comprehensive Definition of Technical Communication, *Journal of Technical Writing and Communication, 21*:2, pp. 133-153, 1991.
16. C. P. Snow, The Two Cultures, *New Statesman*, pp. 413-414, October 6, 1956.
17. J. Dewey, *The Public and Its Problems*, Swallow Press, Athens, Ohio, 1985.
18. J. Ellul, *The Technological Society*, John Wilkinson (trans.), Knopf Books, New York, 1964.
19. L. Winner, *Autonomous Technology*, Massachusetts Institute of Technology Press, Cambridge, Massachusetts, 1977.
20. M. D. Whitburn, The Ideal Orator and Literary Critic as Technical Communicators: An Emerging Revolution in English Departments, in *Essays on Classical Rhetoric and Modern Discourse*, R. J. Connors, L. S. Ede, and A. A. Lunsford (eds.), Southern Illinois University Press, Carbondale, Illinois, 1984.

21. K. B. LeFevre, *Invention as a Social Act*, Southern Illinois University Press, Carbondale, Illinois, 1987.

22. C. Bazerman, *Shaping Written Knowledge: The Genre and Activity of the Experimental Article in Science*, University of Wisconsin Press, Madison, Wisconsin, 1988.

23. D. Sullivan, Political-Ethical Implications of Defining Technical Communication as a Practice, *Journal of Advanced Composition, 10*:2, pp. 375-386, 1990.

24. D. B. Bradford, The New Role of Technical Communicators, *Technical Communication, 32*:1, pp. 13-15, 1985.

25. G. M. Parsons, Ethical Factors Influencing Curriculum Design and Instruction in Technical Communication, *IEEE Transactions on Professional Communication, 30*:3, pp. 202-207, 1987.

26. E. Harris, In Defense of the Liberal-Arts Approach to Technical Writing, *College English, 44*:6, pp. 628-636, 1982.

27. R. Rutter, Poetry, Imagination, and Technical Writing, *College English, 47*:7, pp. 698-712, 1985.

28. M. S. Samuels, Technical Writing and the Recreation of Reality, *Journal of Technical Writing and Communication, 15*:1, pp. 3-13, 1985.

29. A. Lyon, Paideia to Pedantry: The Dissolving Relationship of the Humanities and Society, *Journal of Technical Writing and Communication, 18*:1, pp. 55-62, 1988.

30. A. D. Manning, Literary vs. Technical Writing: Substitutes vs. Standards for Reality, *Journal of Technical Writing and Communication, 18*:3, pp. 241-262, 1988.

31. P. M. Dombrowski, Challenger and the Social Contingency of Meaning: Two Lessons for the Technical Communication Classroom, *Technical Communication Quarterly, 1*:3, pp. 73-86, 1992.

32. W. B. Weimer, Science as a Rhetorical Transaction: Toward a Nonjustificational Conception of Rhetoric, *Philosophy and Rhetoric, 10*:1, pp. 1-29, 1977.

33. L. Fleck, *Genesis and Development of a Scientific Fact*, F. Bradley and T. J. Treem (trans.), University of Chicago Press, Chicago, Illinois, 1979.

34. T. Kuhn, *The Structure of Scientific Revolutions*, University of Chicago Press, Chicago, Illinois, 1970.

35. P. Feyerabend, *Against Method*, Verso Books, New York, 1988.

# CHAPTER 2

# Rhetoric of Science

The rhetoric of science is a fertile new field of study. It might, however, appear perplexing to traditionalists. Indeed, the concept of the rhetoric of science itself appears to be problematic in several ways discussed below. As we will see, these problems stem from assumptions that many contemporary thinkers are challenging. In this chapter, I will first present an overview of the rhetoric of science by considering the questions, Rhetoric *and* Science? and Rhetoric *or* Science?, followed by a three-level classification scheme. These two questions encapsulate the long history of debate about the relation of rhetoric to science. This is followed by a literature review, then an introduction to and critical commentary on the two reprinted articles.[1]

## RHETORIC *and* SCIENCE?

Both the terms "rhetoric" and "science" have their own difficulties of definition, as R. A. Harris explains [1]. Rhetoric, he says, has a variety of definitions and both noble and ignoble senses (or "eulogistic" and "dyslogistic" senses, as Simons elsewhere puts it [2]). Science, too, has a variety of definitions and senses. The rhetorician R. Weaver even goes so far as to say "science" has taken on a religious sanctity and has become a "god-term" of our culture [3]. Naturally, then, the joining of these two indefinite terms has a compounded indefiniteness.

---

[1] As mentioned earlier, I will generally take "science" and "technology" as interchangeable, particularly in not distinguishing between technical and scientific communication unless circumstances warrant a distinction. This chapter deals with the rhetoric of science; the rhetoric of technology is a new-born area yet to be defined clearly. See Harris's *Rhetoric of Science* for the few sources in the rhetoric of technology [1].

## SCIENCE *or* RHETORIC?

Traditionally, science has been defined as specifically separate from, even opposed to, rhetoric. Science, it has been held, deals with what can be empirically demonstrated to be true regardless of our expectations, beliefs, or opinions. Such truth has been held to be absolute—certain, immutable, objective, and without contingency. Science also deals with the interrelation of apodictic (certain) knowledge, for instance in deductively deriving new knowledge from already-known truths. One of the most famous (although failed) intellectual movements of the early twentieth century, logical positivism, specifically aimed at finding and working with only what was known certainly, positively. It pursued the ambitious project of superseding the fuzziness of word-based language with the exact language of mathematical logic operating on positively known facts. The project failed in its original intent but succeeded wonderfully in revealing that nothing can be known absolutely. It is no coincidence that Einstein's relativity, Planck's probabilistic quantum physics, and Heisenberg's indeterminacy principle were developed in and after the period of positivism.

Rhetoric, it has conversely been held, deals with what cannot be absolutely true: opinion, probability (not statistical but conjectural), and social contingency. As Aristotle pointed out, only rhetoric and its sister art dialectic are indeterminate in that they can "draw opposite conclusions impartially." Scientific reasoning and deductive logic, on the other hand, can draw only single, binding conclusions.

From this traditional understanding of rhetoric and science, the term "the rhetoric of science" is an oxymoron, a mixing of apples and oranges with no elements in common. The contemporary understanding of the issue is quite otherwise, however, and denies the validity of the definitions on which the above reasoning is premised.

Science, as we will see repeatedly, is now often understood to be only a particular form of opinion, consensually negotiated and ratified against communal standards. Though scientific standards are usually quite rigorous, they are nonetheless shifting and amenable to alteration.

Rhetoric, too, is not now understood as it had been. Rhetoric to contemporary rhetoricians includes rigorous, deductive demonstration such as used in mathematics as much as it includes appeals to emotions. Even the historical distinction between deduction and induction has become muddled.

In addition, careful users of the term "rhetoric" use it only in the noble sense, not in the ignoble sense of deception and manipulation. Rhetoric for us involves full, open debate among mutually respecting peers, the clash (*agon*) of the two opposing arguments allowing the inherent truth of the matter to reveal itself.

This natural assertion of the truth is the nexus between rhetoric and science. The clash of arguments occasioned by the introduction of a new theory, for example, allows the natural truth (however contingent) of one theory to assert itself over the other. Such assertion in science often involves executing a careful experiment to test competing theories. The result might, of course, be that neither is vindicated in an either-or manner but that both are reinterpreted (e.g., Newtonian mechanics as only a special case of Einsteinian mechanics).

This new understanding of "rhetoric" and "science" not only makes it meaningful to conjoin the two terms, it compels the conjoining. Understanding science as a special form of argued persuasion within a particular community, that is, compels us to examine science from the point of view of rhetoric. This is not necessarily to reduce science to rhetoric (though some do, e.g., Gross's reprint) but only to understand the workings of science in this particular way. Rhetoric of science, in a way, is even itself rather scientific in searching for the non-obvious features of phenomena and then reflexively using these features to analyze still other phenomena. In this way, new insights are gained without necessarily diminishing the phenomenon studied, in this case, science.

## THREE LEVELS OF RHETORIC OF SCIENCE

The rhetoric of science can be readily understood from a three-level classification scheme. This scheme is systematic, parsimonious, comprehensive of key theoretical distinctions, and robust in highlighting the complex, subtle aspects of this topic. It is offered only as an alternative to other schemes (e.g., Harris's six-part system), not as a replacement.

### Level One: Names and Concepts

The most fundamental but least obvious level of rhetoricity involves names and concepts such as phlogiston, force, and quark. Due to habit and to the preeminence science has had in our culture, we are not used to thinking of denomination as an act of rhetoric. We instead think that what is named actually exists as itself prior to and separate from our conceptualization of it and the act of naming. A tree, we are used to thinking, is a tree regardless of what we call it and regardless of whether anyone is around to hear it fall. In like manner, we often think of concepts such as quarks as being real things and our use of "quark" as an instance of precise, concrete, and natural referentiality.

This thing-based referentiality has a long though checkered history in language studies, but its history relating to science is particularly germane.[2] Many articles

---

[2] Theorists of language such as Saussure and Wittgenstein explain that our perceived reality is constituted, partly if not wholly, by our language. For them, words do not refer to pre-existent things but to other words and systems of words and to the concepts they represent. See for instance C. B. Guigon, the philosopher of science, for a concise review of self-reflexive, self-interpreting language games, one of which is Science [4].

on the humanistic aspects of technical communication have cited, for example, the founders of the Royal Academy of Science as banishing rhetoric from science so as to purify scientific language. S. M. Halloran and M. D. Whitburn, for instance, cite Sprat's famous dictum to treat a number of things in an equal number of words [5]. C. R. Miller, too, cites the positivistic attempts of A. N. Whitehead and B. Russell to represent the facts of science in the nonlinguistic symbolism of mathematics [6].

Material referentiality, however, is not always what it is held to be—concrete and tangible. It instead, surprisingly, can be understood as an artifact of other, diametrically different historical legacies, idealism and essentialism. Oftentimes scientific and technical referentiality looks beyond the particular qualities of particular objects to the thing or phenomenon as it supposedly truly is prior to and separate from our perceptions and cognitions. That is, referentiality looks beyond the particulars to the general, from a particular apple falling from a particular tree at a particular time to a general theory of gravitation applying to all apples and to all the stars and planets as well. [3]

Very often we think that the things of observation exist before or prior to our words indicating them. From a contemporary point of view, however, such things are already reflections of what was sought because we find only what we are capable of and expect to see. P. Feyerabend makes this point very strikingly in his historical study of Galileo's observations of Jupiter's moons [10]. Galileo saw lights alongside Jupiter as moons orbiting that planet only because he first entertained the notion of heliocentrism, which conditioned his cognitions and perceptions. Thus, one's prior conceptualizations condition and make possible one's observation.

Phlogiston is a good example of the social contingency of things. In classical times, the presence of a substance called phlogiston was the explanation for fire and heat, a concept accepted throughout that culture as authoritative. This concept "worked" for the purposes of that culture. During the rise of modern science, however, phlogiston was sought empirically but never found and, more importantly, was revealed to be unnecessary and incompatible with the caloric, kinetic view of heat. That is, within the scientific culture of the seventeenth and eighteenth centuries, the concept of phlogiston did not "work" and so was discarded. [4] Thus

---

[3] The social constructedness of scientific referents is being researched intensively. A. Gross, for instance, explores scientific language as a "myth" which pretends to refer to nature but which really only returns us reflexively (though inapparently) to our own constructed system of referents [7].

[4] C. Bazerman's chapter *How Language Realizes the Work of Science* deals with rhetoric in this way, that is, as seeing language not as referring to already-existing knowledge but as constituting knowledge [11]. From Bazerman's perspective, science is nothing but a semiotic system. In my following chapter on social constructionism, however, we will see that Bazerman maintains material referentiality and empiricism as essential to science and so limits the rhetorical constructedness of scientific knowledge.

it is an example of a scientific thing that "existed" then ceased to exist, becoming a non-entity, on the basis of later developments.

The obvious objection to the example of phlogiston is that terms such as "phlogiston" have a referent only within a particular theory while other terms such as "house" refer to objects in the real world separate from theories. Rhetoricians, however, contend that such realistic terms as "house" are nonetheless theory bound, though the theoretical underpinning is obscured in common usage (such obscurations are also noted by social constructionists—see Chapter 3). One may view a house from a variety of theoretical perspectives: aesthetically for its beauty, architecturally for its soundness or style, capitalistically as personal property, communistically as communal property, militarily as a defensive position, entomologically as a feast for termites, chemically as a collection of elements and compounds, physically as matter potentially convertible to energy, or domestically as a home.

Indeed, the rhetoricity of apparently real referents is a point of major intellectual contention. The 1989 conference of the International Society for the History of Rhetoric at Johns Hopkins, for example, featured a panel debate between prominent rhetoricians of science and traditional philosophers of science. The traditional philosophers of science insisted on the reality of the electron irrespective of theoretical or rhetorical constructions, while the rhetoricians of science (including A. Gross and C. R. Miller, both with articles reprinted in this collection) insisted on the theoretical and therefore rhetorical contingency of the concept and on the ultimate obscurity of the thing-in-itself.

## Level Two: The Ratification of Scientific Theories

The intermediate and most obviously rhetorical level in which rhetoric is engaged in science is that of the continual evolution of scientific knowledge and the negotiation, advocacy, and debate which occasions this evolution. A good illustration of this clear rhetoricity is the famous theory of paradigm shifts found in T. Kuhn's landmark work, *The Structure of Scientific Revolutions* [12]. Another good illustration is Halloran's rhetorical history of the rejection and acceptance of equivalent representations of the structure of the DNA molecule, the acceptance of the later Watson and Crick model being conditional on having been more strenuously argued by more well-known proponents to a audience more receptive to the idea [13]. Gross's book, *The Rhetoric of Science* also contains many thoroughly researched instances [14].

Because most people already understand that acceptance of a theory by a scientific community necessitates argumentation and demonstration, which are granted forms of rhetoric, most readers will have no difficulty accepting that science is rhetorical at least at this level.

## Level Three: Scientific over Other Knowledge

Rhetoric is also involved in the enterprise of science in establishing (usually covertly) the primacy of scientific knowledge over other forms of knowledge. This primacy includes vitally important political and economic decisions about, for instance, which potential research topics will be pursued rather than others, or how much in the way of resources will be allocated to such research in a context of tightly constrained resources.

C. Waddell, for example, has studied the rhetoric of science policy, which is concerned with the role of rhetoric in shaping science policy decisions such as what sorts of scientific research will be permitted and which will be banned in particular communities such as Cambridge, Massachusetts [15]. Likewise, decisions by governmental scientific agencies about how much federal money will be awarded to the search for indications of extra-terrestrial intelligence are made in the context of both competing scientific research projects and competing non-scientific projects such as providing good medical care for all American children. In a decision such as this, the intrinsic value of basic scientific research is continually argued, defined, and re-defined, either implicitly or explicitly.

This level of rhetoricity is less apparent than the middle level but no less important. Indeed, many critics of science, including ethicists, feminists, and philosophers of science, take science to task at precisely this level (see also Chapter 5, Ethics and Chapter 6, Feminist Critiques of Science). These critics argue against the very idea that scientific knowledge, due to its objectivity and unemotionalness, should be privileged as somehow better, truer, and more valuable than other sorts of knowledge. Feminists such as C. Gilligan, M. J. Larrabee, and N. Noddings argue that compassionate caring is at least as important a basis for "knowledge" as the cold dispassion of science. Advocates for relaxing FDA standards for medications such as AZT to treat AIDS make similar arguments, holding that compassion is precisely the issue, not rigorous scientific proof of efficacy.

More incisively, ethicists such as J. Ellul and S. Monsma argue that the dispassionate disinterestedness and object-orientation of science do not indicate that science transcends moral values but that these features are themselves values taken quite seriously and moralistically (see Chapter 5, Ethics). These values must be comprehended on a par with other values and must be grappled with in the arena of ethics. To assume that science is the last word on knowledge, they argue, is a grave mistake.

## LITERATURE REVIEW

S. M. Halloran's *Technical Writing and the Rhetoric of Science*, though early, provides an excellent synopsis of the relation between rhetoric and science [16]. Halloran traces the historical dualistic separation of them promulgated by Aristotle followed by the contemporary conjoining of them, citing the example of the publication of Watson and Crick's research on the structure of DNA. One of the ramifications of recognizing the role of rhetoric in science, Halloran points out, is also to recognize the complex role of the character of the scientist as a member of a scientific community and to role of science as an activity embodying a system of values sanctioned by our society.

D. Journet's article *Rhetoric and Sociobiology* deals with rhetoric in fairly general terms [17]. Journet establishes that science is rhetorical using the example of the debate over E. O. Wilson's controversial sociobiology as a legitimate scientific theory. This rhetorical debate focused not on theoretical substance or even on empirical evidence but on the "political, ethical, and moral ramifications" of its application to humans [17, p. 339]. Though this debate had more to do with science policy than with science per se and occurred more in popular forums than scientific ones, Journet's article can usefully serve as an introduction to the rhetoric of science. [5]

## R. Rutter

Rutter's *History, Rhetoric, and Humanism: Toward a More Comprehensive Definition of Technical Communication* deals little with the rhetoric of science specifically [19]. It instead calls for the need for interest and education in rhetoric in contemporary society to counterbalance our society's lopsided and potentially harmful preoccupation with science and technology. Rutter argues that technical communicators should be rhetoricians politically engaged, socially responsible, and constructively critical of the scientific and technical knowledge they communicate.

## J. P. Zappen

Zappen's domain of interest in *A Rhetoric for Research in Sciences and Technology* is rather circumscribed, the research practice of science and technology from the point of view of the context-oriented argumentation theory of S. Toulmin, a sort of theory of rhetoric [20].

One of Zappen's key findings is that research in the sciences is not radically different from research in the technologies, as is often assumed, and is highly socially contingent. Both scientific and technological research reports reflect common concerns about goals, current capacities, problems, solutions, and criteria of evaluation. Zappen also finds that Toulmin's argumentation theory is an

---

[5] Another article by Journet, *Writing, Rhetoric, and the Social Construction of Scientific Knowledge*, is principally about social constructionism [18]. Because rhetoric of science and social constructionism are so closely allied, the reader interested in rhetoric of science might find this article useful though *Rhetoric and Sociobiology* is deeper in this area.

important tool for investigating and defining the details of the contexts operative in research reports.

J. P. Zappen in *Historical Studies in the Rhetoric of Science and Technology* points out that though early studies in the rhetoric of science and technology emphasized differences between classical rhetoric and the rhetoric of contemporary science and technology, more recent studies emphasize the similarities such as concern about the role of style and arrangement [21]. The political and ethical dimensions of scientific discourse are also explored in recent research. Indicating new directions in rhetoric of science and technology, Zappen cites B. Kimball to explain that civic or community-grounded rhetorics such as Isocrates's and Cicero's can counterbalance the historical predominance of philosophical rhetorics such as Plato's and Aristotles's. Zappen also indicates, citing S. Harding, that the feminist critique of science is another direction with important political and ethical implications.

## C. R. Miller

Miller's *What's Practical About Technical Writing*, surprisingly, is largely about rhetoric and this surprise is precisely her point [22]. "Practical" regarding rhetoric frequently means to use a handbook or a bag of tricks (*techne*). Miller explains, however, that an equally valid and more important meaning concerns practical wisdom or prudential reasoning (*phronesis*) and involves heavily both ethics and rhetoric. From this expanded Aristotelian perspective, rhetoric in technical communication is both highly practical and highly conscious of social responsibilities.[6]

## B. Herzberg

Herzberg's *Rhetoric Unbound: Discourse, Community, and Knowledge*, traces the history of the troubled relation between rhetoric and epistemology (the study of the bases of knowledge) from sixth century B.C. Greece to our postmodern times [24]. Herzberg explains that both the earliest and the latest thinkers conceive of knowledge as inseparable from rhetoric. For Herzberg, the rhetoric of science is a social and cultural critique.

In our own times, M. Foucault, the literary theorist and culture critic, Herzberg says, brings the history full circle in his reaffirmation of the insights of the Sophists that discourse embodies and constitutes knowledge.

---

[6] For a deeper exploration of the concept of practical wisdom as it relates to rhetoric and to pragmatism, see J. A. Mackin, Jr.'s *Rhetoric, Pragmatism, and Practical Wisdom* [23].

## R. Allen Harris

Harris's *Rhetoric of Science* offers a lucid explanation of the role of rhetoric in science, a systematic categorization of the widely-varied discussions in this area, and a comprehensive literature review [1]. (His other article on rhetoric of science is reprinted in this volume.) It is concerned less with describing how science is rhetorical, more with exploring how "rhetoric" is used with respect to science and with differentiating rhetoric from other intellectual perspectives on science.

Harris identifies six general categories of rhetoric of science: rhetoric of technology; rhetoric of religion in science; rhetoric of scientific composition; rhetoric of scientific language; rhetoric of public science policy; and prototypical rhetoric of science. This categorization reflects, Harris feels, the natural articulation of this subject, any other categorization being forced, unfitting, and unnatural. In addition, rhetoric of science has two senses, empirical and theoretical. Harris develops a prototype to definition of rhetoric of science as argumentation about the data and theory of science.

## L. J. Prelli

Prelli's *A Rhetoric of Science: Inventing Scientific Discourse* is unique in considering the rhetoric of science less as logic and argument, more as rhetorical "invention" [26]. Prelli reviews developments in the sociology of science to show that scientific knowledge is neither strictly "discovered" nor "found" but socially constructed and negotiated, then explores how science is done by conceiving, constructing, and presenting scientific claims so as to achieve acceptance among other scientists.

Prelli takes a modest position on the rhetoricity of science, seeking only the rhetorical principles on which scientific discourse is created and judged. This position differs from that of Gross, who, we will see later, takes a radical position holding that science is nothing other than rhetoric.

The key rhetorical aspects of scientific discourse for Prelli are its grounding in the situation and audience, its reasonableness, and its social inventedness, its lines of argument (rhetoric topics or *topoi*), and its rhetorical issues (*stases*). Prelli, though largely classical, takes an eclectic approach to rhetoric spanning a broad variety of theories.

## H. W. Simons

Simons's *Rhetoric in the Human Sciences* (1989), which he edited, contrasts the human and the natural sciences and parallels his other book, discussed next [2, 27]. In this collection, Simons emphasizes the interpretive and patently self-reflexive natures of the human sciences which makes them particularly susceptible to rhetorical treatment. [7]

Simons's other book, *The Rhetorical Turn: Invention and Persuasion in the Conduct of Inquiry*, defines a field of study almost congruent with the rhetoric of science: the rhetoric of inquiry [31]. The rhetoric of inquiry views rhetoric not only as advocacy and persuasion but also as a manner of inquiry, that is, in the way we conceptualize, examine, and develop knowledge about the world. With Simons's conceptualization, the duality between science and rhetoric becomes muddled as discovery shades into (rhetorical) invention and explanation shades into (rhetorically contextualized) interpretation.[8]

Simons points out the need to move discussions about the rhetoricity of science away from a simple and marginally fruitful either-or dichotomization: either science is only rhetoric or it is essentially other-than rhetoric. Simons explains that we need to recognize that the fruitfulness of clashes of the either-or sort lies in the potential for a new, different conceptualization that is neither one nor the other position. Simons suggests we should discuss less whether a rhetorical account of science is superior to an empiricist account of science, more how rhetoric can serve as a useful adjunct to conventional science.

Particularly interesting is Simons's exploration of the limitations of an extreme position on the rhetoricity of science (viz., that science is all and nothing but rhetoric). Two difficulties with this extreme position are explored in several of the essays not by Simons. One difficulty is that if rhetoric of science is closely allied with the pervasive antifoundationalism in academia, it would be inconsistent in the same intellectual movement to assert that rhetoric is foundational to science. The other difficulty is that the extreme rhetoric of science position is strongly suggestive of radical relativism. A radical relativist position in effect throws knowledge up for grabs by denying preference to any particular sort of knowledge, and almost perversely renounces the unquestionably significant advances in knowledge made by science in the last five hundred years. Thus an extreme rhetoric of science position risks evaporating the object it is supposed to illuminate, science, by reductively substituting rhetoric for all of science.

---

[7] W. Dilthey was one of the principal exponents of the notion of the human sciences (*Geistwissenshaften*) as essentially different from the natural sciences (*Naturwissenshaften*). Dilthey thus undermined attempts to make the human sciences as rigorously deterministic as the natural sciences; in so doing, he helped to characterize and define the limitations of positivism.

[8] For contemporary references on interpretation regarding the sciences, I recommend the Introduction in D. R. Hiley et al.'s *The Interpretive Turn* and chapters by H. L. Dreyfus and J. Rouse in Hiley [28-30].

## J. S. Nelson and A. Megill

Nelson and Megill's *The Rhetoric of the Human Sciences: Language and Argument in Scholarship and Public Affairs*, compiles essays from a wide range of authors in this area [32]. Nelson and Megill's separate article, *Rhetoric of Inquiry: Projects and Prospects*, can serve as a precis of their book [33]. The thrust of their work coincides generally with Simons's and so will not be reviewed in detail here, except for the ties to postmodernism.

Nelson and Megill trace three intellectual strains which historically have led to the rhetoric of inquiry: antifoundationalism, the reconceptualization of science, and the identification of the rhetoricity of science. The reader might find particularly interesting the political and culture criticism in Nelson and Megill. They characterize the "absolutized dichotomy" between truth and opinion as an expression not only of some strains of classical and Renaissance thought but also of modernism itself. Reactions against and correctives to modernism, under the name "postmodernism," result from the three skeins of contemporary thinking just mentioned. For Nelson and Megill, the rhetorics of science and inquiry are expressions of postmodernism, especially deconstructionism stemming from Foucault and Derrida and the historicism of White and Gadamer.

## A. G. Gross

Gross's book, *The Rhetoric of Science*, is perhaps the most authoritative work on the rhetoric of science to date [14]. It acts as an intellectual link between the rhetoric of science and social constructionism, the topic of the next chapter.

This book of Gross's own essays duplicates and elaborates his articles, including the one reprinted in this volume. The epilogue, however, is not represented elsewhere and is important as the single most explicit and definitive statement his methodology, his purposes, and the general thrust of his work. This statement in turn is important because otherwise many scientists and philosophers of science might have little about which to argue with Gross. Gross strongly argues against many of tenets of the conventional understanding of science, in particular the idea that science genuinely "refers" to a mind-independent reality.

The apparent independence of this reality, Gross explains, is only a presupposition entirely contingent on one's prior acceptance of the absolute authority of Science, and therefore on a belief. Therefore the "reality" studied by science is *not* independent at all and is indeed dependent on and constituted by the very enterprise of science.

Gross also challenges the claim that science is a process of successive approximation, an iterative refinement converging on absolute reality. This convergence claim holds that successive theories such as Aristotle's physics and Newton's physics say roughly the same thing, the latter being a refinement of the former. Gross explains, however, that the true history of science is not a progression of refinements converging on absolute success but rather a conglomerate of different understanding, many of which are "incommensurable" (a technical term from the history and philosophy of science). Gross constructively offers, in place of traditional philosophical realism, a realism based on the rhetorical nature of facts. From this perspective, science is only a system of beliefs, one among many similarly valid, defensible, and viable systems.

Gross's article *Rhetoric of Science Without Constraints* is a succinct statement of the radical or "strong" position on the rhetoric of science (also discussed by Rude below) [34]. For Gross, even the "brute facts of nature" are artifacts of rhetorical negotiations. Thus there are no constraints on the application of rhetoric to science—science is nothing but rhetoric. My critical comments regarding the reprinted article apply to this article as well.

Gross's refutation of the special authority of science has three parts based respectively on prediction, empirical regularities, and the indirectness of observations. Gross also debunks the notion of the "recalcitrance" of nature as the root which distinguishes science from rhetoric. This recalcitrance, Gross says, cannot be independently characterized in any way that is not innately rhetorical and therefore cannot of itself demonstrate that science is anything other than rhetoric.

Gross's all-or-none position seems so narrow and extreme as almost to guarantee contention and needlessly to tempt refutation. A Rortian frame of reference, for instance, itself strongly social constructionist and anti-representationalist, does not necessitate an absolute rejection of any validity to science. Rorty argues, rather, for an alteration of both the assumptions and the goals of our search for knowledge. Regarding these assumptions, he challenges the view that the nature of material reality itself compels the methods of scientific investigation, holding instead that these methods are always already our own productions and have authority only by virtue of our choosing to take them as having that authority. Thus Rorty allows both (relative and conditional) separation and peaceable co-existence among rhetoric, social constructionism, pragmatism, and science, albeit on reconceptualized bases.

## C. D. Rude

Rude's *The Rhetoric of Scientific Inquiry* draws parallels between the radical stances of Gross's *The Rhetoric of Science* and B. Latour's (a philosopher of science) *Science in Action: How to Follow Scientists and Engineers Through Society* [35].[9] Their position basically is that scientific writing is nothing other

---

[9] To be sure, most readers conceive of Latour as writing about the sociology of science and scientific knowledge, not about the rhetoric of science. Nonetheless, Latour does discuss the social/rhetorical context and persuasive argumentation of scientific discourse and so legitimately can be understood as dealing with the rhetoric of science as well.

than rhetorical statements. She finds their position highly congruent and both highly divergent from more moderate stances such as those of Bazerman and Kuhn which retain the notion of traditional notion of facticity.

## REPRINTED ARTICLES

### R. A. Harris

Harris's *Assent, Dissent, and the Rhetoric of Science*, is important in three ways [25]. First, Harris clarifies how persuasion and rhetoric are involved in the enterprise of science by exploring the history of several important scientific debates over theory. This exploration also shows that there is no sort of persuasion peculiar to science. Rhetoric operates in the scientific arena just as it does in other arenas, though the specific subjects are of a scientific nature.

Second, Harris shows the connection between "truth" (supposedly characteristic of science) and "opinion" (supposedly characteristic of rhetoric) in scientific debates and the non-obvious contextually contingent usage of these terms. Harris goes so far as to show that what we know as truth and opinion are only different representations of the same activity and the same knowledge. Thus debate as the spirited, messy indeterminacy of contention is absolutely vital to science, its *sine qua non*. This is "opinion" in its wholesome sense, of course, critically examined and socially ratified among peers. Without progressive challenges from, and subsequent accommodations to and assimilation of, new theories, science become ossified, as good as dead. Then it becomes only the uncritically passing on of received knowledge from entrenched authority figures, "opinion" in the pejorative sense of dogma.

Third, Harris explores the very human psychology by which science confronts change. He points out that the cognitive dissonance of debate, notwithstanding its ultimately beneficial outcome, necessarily means that it is difficult, painful, and genuinely trying. It should not be surprising, then, that debate is avoided. It is simply easier to accept the current state of theoretical affairs as absolutely and incontestably true and conversely to take theoretical challengers as only insubstantial expressions of mere opinion. Thus the debates occasioned by novel theories appear to be abnormal, even sophistic in seeming to make the weaker case the stronger. Thus dissension *seems* contrary to fact and apparently untrue. Only after a strenuous uphill battle can new theories come to prevail over old in a sort of evolutionary selection process. Therefore new theories must not be discredited simply because they radically clash with what is accepted as obviously true.

In these three ways, Harris shows that rhetoric not only permeates science but is inextricably essential to it.

## A. G. Gross

Gross's article, *Discourse on Method: The Rhetorical Analysis of Scientific Texts*, sketches many of the issues elaborated in his other works and so serves as an encapsulation of his key points [36]. Chief of these is the radical or "strong" position on the rhetoricity of science that minimizes if not rejects the notion of facts. Facts, if they ultimately exist at all, have no meaning of themselves and do not constitute science. Science, Gross says, is knowledge which is equivalent to meaning, and because only statements (not facts per se) can "mean," then science is persuasion. "[O]nly through persuasion are importance and meaning established"; therefore, only rhetoric can reveal what science truly is [36, p. 170].

Gross explicates science using the rhetorical elements of stasis theory; rhetorical invention, topics, and genres; and the rhetorical appeals of *ethos, pathos,* and the readily-granted *logos.* In addition, he argues that if differing preconceptions and theoretical expectations yield wildly differing "sciences" with equal claims to the authority and privilege that that term carries, then there is no one, absolute science, only the shifting, historically conditioned significations attached to that term—and only rhetoric can account for how various significations come to be accepted within a given scientific community. As a corollary, the universe as represented in science is not immutable, as claimed, but is as mutable as these various representations.

It is easy to read Gross in a pejorative way as radically diminishing science, reducing it to rhetoric. A more constructive reading (and perhaps truer to Gross's intent) construes the article as an expansion and elevation of rhetoric. That is, implicit in Gross's position is a view of rhetoric as the fundamental activity of our humanity and our quest for knowledge. For Gross as for other rhetoricians such as K. Burke or M. Leff, our knowledge, our world, even our sense of ourselves is continually negotiated, revised, and even constituted through our language. Thus Gross does not so much impugn science as being only "mere rhetoric" as counter the historical "degradation" of rhetoric [36, p. 183].

Another critic voicing similar concerns and arguing for the elevation of rhetoric is W. B. Weimer in *Science as a Rhetorical Transaction: Toward a Non-Justificational Conception of Rhetoric* [37]. Weimer explains that all knowledge is cognitive conceptualizations and constructions imposed on innately ambiguous, indeterminate entities. All scientific explanation is therefore never essentially justified but are only argued conceptualizations which the audience is convinced of or not on the basis of socially contingent warrants. An additional consonant voice is S. M. Halloran's in *Rhetoric in the American College Curriculum: The Decline of Public Discourse* which traces the diminishment of rhetoric studies in American higher education until recent decades [38].

Allied with Gross's elevated view of rhetoric is his larger purpose, not to reject science so much as to reunite the sciences and the humanities, to mend "the rupture caused by the dualism of Descartes" [36, p. 183]. "Because it sees science wholly as a product of human interaction, rhetoric of science is a gesture in the direction of such reconciliation [between the sciences and the humanities], an argument for the permanent bond that must exist between science and human needs" [36, p. 183]. Thus, Gross's work returns the human element to science as he argues for humanity in science (by revealing the humanistic side of science).

By way of criticism, it seems to me that Gross undertakes a project perhaps both too adversarial and too extreme. I do not disagree with any of Gross's analyses of scientific texts or histories, only with the scope of the conclusions he draws from them. Revealing the rhetorical nature of science and scientific texts does not necessarily entail the repudiation or invalidation of some of the aims of science. That is, asserting the rhetoricity of science need not entail denying a mind-independent material world which exists somehow separate from our beliefs.

There is also a problem of self-exclusion. It is one thing to undermine the absolute privilege and authority historically accorded science, another thing to assert that a similar absolute privilege and authority should be accorded another conceptual framework. In effect, Gross seems to exclude rhetoric from the same arguments with which he attacks science.

The critical reader is thus left with a fundamental unanswered question: In the project of disprivileging science and disabusing us of received notions about the absoluteness and primacy of science and scientific knowledge, is it necessary to elevate rhetoric to a parallel privilege? Considering Gross's own criticism of facts, we are left wondering why the "fact" of the rhetoricity of science should not be as problematic as the facts of scientific realism.

One of my reasons for these criticisms has to do with the audience of technical communication and our accommodation of them. By couching the rhetoric of science position in an adversarial and extreme form, we risk needlessly alienating our audience of scientists, technicians, and engineers. A more accommodating and humanistic approach seeks to meet our audience on their own terms or at least on some common, middle ground.

A middle ground position avoids the extremes just mentioned. In his closing remarks, Gross seems to say that rhetoric subsumes science and that science is *only* and nothing but rhetoric. A pluralistic middle ground position, on the other hand, would acknowledge that both rhetoric and science have their own sorts of knowledge and spheres of application, overlapping in many ways yet distinct in others. This overlap would permit acknowledging the rhetoricity of science and scientific texts without reducing them to nothing but rhetoric. Likewise, rhetoric could be understood as sometimes itself scientific (as in the rhetorical theories of Campbell, Bain, Richards, and Kinneavy), at other times as using scientific information for rhetorical ends such as in arguments about science policy. Thus we can acknowledge the role of rhetoric in constituting science without co-opting science to rhetoric.[10] This simultaneous overlapping yet distinctness might be something like the interrelation between psychotherapy and biochemistry, both interested in human being but from distinctly different frames of understanding, interest, and meaning.

Thus rhetoric and science form a graded continuum. One way of looking at (and thus constituting) a world *highlights* the rhetorical while another way *highlights* the facticity of science, with a quite different collection of values, principles, and goals. Neither of these two ends of the continuum is radically, absolutely separate from the other. As a convenience, however, and as a useful approach for particular circumstances, cultures, and purposes, the two might usefully be treated (at least provisionally) as separate.

## REFERENCES

1. R. A. Harris, Rhetoric of Science, *College English, 53*:3, pp. 282-307, 1991.
2. H. W. Simons (ed.), *Rhetoric in the Human Sciences*, Sage, Newbury Park, California, 1989.
3. R. Weaver, *Language is Sermonic*, R. L. Johannesen, R. Strickland, and R. T. Eubanks (eds.), Louisiana State University, Baton Rouge, Louisiana, 1970.
4. C. B. Guigon, Pragmatism or Hermeneutics? Epistemology after Foundationalism, in *The Interpretive Turn*, D. Hiley, J. F. Bohman, and R. Shusterman (eds.), Cornell University Press, Ithaca, New York, 1991.
5. S. M. Halloran and M. D. Whitburn, Ciceronian Rhetoric and the Rise of Science: The Plain Style Reconsidered, in *The Rhetorical Tradition and Modern Writing*, J. J. Murphy (ed.), Modern Language Association, New York, 1982.
6. C. R. Miller, A Humanistic Rationale for Technical Writing, *College English, 40*:6, pp. 610-617, 1979.
7. A. G. Gross, The Origin of Species: Evolutionary Taxonomy as an Example of the Rhetoric of Science, in *The Rhetorical Turn: Invention and Persuasion in the Conduct of Inquiry*, H. W. Simons (ed.), University of Chicago Press, Chicago, 1990.
8. B. Kimball, *Orators & Philosophers*, Teacher College of Columbia University, New York, 1986.
9. R. Rorty, *Objectivity, Relativism, and Truth*, Cambridge University Press, Cambridge, England, 1991.
10. P. Feyerabend, *Against Method*, Verso Books, New York, 1988.
11. C. Bazerman, How Language Realizes the Work of Science, in *Shaping Written Knowledge: The Genre and Activity of the Experimental Article in Science*, University of Wisconsin Press, Madison, Wisconsin, 1988.

[10] This mutual toleration of rhetoric *and* science also avoids a tautology. If any and every activity is subsumed under rhetoric and treated as nothing other than rhetoric, then what is rhetoric? If all A and all not-A are B, then everything is B, and so denominating something as a case of B tells us nothing about B. Thus an all-subsuming rhetoric could become an uninformative tautology.

12. T. Kuhn, *The Structure of Scientific Revolutions*, University of Chicago Press, Chicago, Illinois, 1970.
13. S. M. Halloran, Technical Writing and the Rhetoric of Science, *Journal of Technical Writing and Communication, 8*:2, pp. 77-88, 1978.
14. A. G. Gross, *The Rhetoric of Science*, Cambridge University Press, London, England, 1990.
15. C. Waddell, The Role of Pathos in the Decision-Making Process: A Study in the Rhetoric of Science Policy, *Quarterly Journal of Speech, 76*:4, pp. 381-400, 1990.
16. S. M. Halloran, Technical Writing and the Rhetoric of Science, *Journal of Technical Writing and Communication, 8*:2, pp. 77-88, 1978.
17. D. Journet, Rhetoric and Sociobiology, *Journal of Technical Writing and Communication, 14*:4, pp. 339-350, 1984.
18. D. Journet, Writing, Rhetoric, and the Social Construction of Scientific Knowledge, *IEEE Transactions on Professional Communication, 33*:4, pp. 162-167, 1990.
19. R. Rutter, History, Rhetoric, and Humanism: Toward a Comprehensive Definition of Technical Communication, *Journal of Technical Writing and Communication 21*:2, pp. 133-153, 1991.
20. J. P. Zappen, A Rhetoric for Research in Sciences and Technologies, in *New Essays in Technical and Scientific Communication: Research, Theory, Practice*, P. V. Anderson, R. J. Brockmann, and C. R. Miller, (eds.), Baywood, Amityville, New York, 1983.
21. J. P. Zappen, Historical Studies in the Rhetoric of Science and Technology, *The Technical Writing Teacher, XIV*:3, pp. 285-298, 1987.
22. C. R. Miller, What's Practical About Technical Writing? in *Technical Writing: Theory and Practice*, B. E. Fearing and W. K. Sparrow (eds.), Modern Language Association, New York, 1989.
23. J. A. Mackin, Jr., Rhetoric, Pragmatism, and Practical Wisdom, in *Rhetoric and Philosophy*, R. A. Cherwitz (ed.), Erlbaum, Hillsdale, New Jersey, 1990.
24. B. Herzberg, Rhetoric Unbound: Discourse, Community, and Knowledge, in *Professional Communication: The Social Perspective*, N. R. Blyler and C. Thralls (eds.), Sage, Newbury Park, California, 1993.
25. R. A. Harris, Assent, Dissent, and Rhetoric of Science, *Rhetoric Society Quarterly, 20*, pp. 105-30, 1990.
26. L. J. Prelli, *A Rhetoric of Science: Inventing Scientific Discourse*, South Carolina University Press, Columbia, South Carolina, 1989.
27. H. W. Simons (ed.), *Rhetorical In The Human Sciences*, Sage, Newbury Park, California, 1989.
28. D. R. Hiley, J. F. Bohman, and R. Shusterman (eds.), *The Interpretive Turn*, Cornell University Press, Ithaca, New York, 1991.
29. H. L. Dreyfus, Heidegger's Hermeneutic Realism, in *The Interpretive Turn*, D. R. Hiley, J. F. Bohman, and R. Shusterman (eds.), Cornell University Press, Ithaca, New York, 1991.
30. J. Rouse, Interpretation in Natural and Human Science, in *The Interpretive Turn*, D. R. Hiley, J. F. Bohman, and R. Shusterman (eds.), Cornell University Press, Ithaca, New York, 1991.

31. H. W. Simons, The Rhetoric of Inquiry as an Intellectual Movement, in *The Rhetorical Turn: Invention and Persuasion in the Conduct of Inquiry*, H. W. Simons (ed.), University of Chicago Press, Chicago, Illinois, 1990.

32. J. S. Nelson and A. Megill (eds.), *The Rhetoric of the Human Sciences: Language and Argument in Scholarship and Public Affairs*, University of Wisconsin Press, Madison, Wisconsin, 1987.

33. J. S. Nelson and A. Megill, Rhetoric of Inquiry: Projects and Prospects, *Quarterly Journal of Speech, 72*, pp. 20-37, 1986.

34. A. G. Gross, The Rhetoric of Science without Constraints, *Rhetorica, IX*:4, pp. 283-299, 1991.

35. C. D. Rude, The Rhetoric of Scientific Inquiry, *IEEE Transactions on Professional Communication, 35*:2, pp. 88-90, 1992.

36. A. G. Gross, Discourse on Method: The Rhetorical Analysis of Scientific Texts, *Pre/Text, 9*:3-4, pp. 169-185, 1988.

37. W. B. Weimer, Science as a Rhetorical Transaction: Toward a Nonjustificational Conception of Rhetoric, *Philosophy and Rhetoric, 10*:1, pp. 1-29, 1977.

38. S. M. Halloran, Rhetoric in the American College Curriculum: The Decline of Public Discourse, *Pre/Text, 3*:3, pp. 245-269, 1982.

# ASSENT, DISSENT, AND RHETORIC IN SCIENCE

## R. Allen Harris

*Lo, a Spartan appears, and says that there never is or ever will be a real art of discourse which is unconnected with the truth.* —Socrates

## 1. INTRODUCTION

Socrates, of course, does not mean to venerate the art of discourse here. He is telling Phaedrus that there is discourse and there is truth. Once you have gone out and dug up the truth somewhere else, you apply the art of discourse to it and fashion a persuasive argument that will permit others to partake also of the truth. Two immediate implications follow from Socrates' position. First, only when the art of discourse, rhetoric, is put to the task of selling truth to the benighted does it become "real." Second, rhetoric is necessary in human affairs just to the extent that humans are unable to apprehend truth directly. It is an unfortunate evil, required because we are rationally degenerate creatures. Both positions have remained very popular over the intervening two millenia. Bitzer, for instance, can still say that "in the best of all possible worlds there would be communication perhaps, but not rhetoric;" we get our truth and knowledge somewhere else, and only our lack of perfection prevents us from casting rhetoric out of the garden [1]. But there is an important lesson in those two millenia that can help us to see the Spartan's words in another light: the sources of truth which rhetoric has been obliged to serve have changed dramatically—from Socrates' dialectic and Aristotle's apodeixis, to Christianity's biblical exegesis and divine revelation, to the current authority on matters of knowledge and truth, Science.

This rotation of leading roles while the supporting actress, Lady Rhetoric, remains constant indicates that the real art of discourse is connected with truth not because of human degeneracy, but because of precisely the reverse, because of our spark of perfection, because we are truth-seeking, knowledge-making creatures who sometimes get it right. We occasionally do something important with rhetoric: we find truth and we build knowledge out of it. When we manage the trick, though, we are so eager to dissociate it from all the foul and inane things we also do with rhetoric that we give the process another name. But these other names are clearly just aliases for rhetoric, or for some subset of rhetorical interests. Dialectic, for instance, is essentially questing debate. Apodeixis is distinguished only by the level of rigor Aristotle demands of the argumentation, not by any qualitative difference. Exegesis is rhetorical analysis. The only possible gap to this pattern is divine revelation, whose capacity to generate truth I will leave to more knowledgeable commentators, pausing only to notice that, true or not, reports of revelation usually involve a fair amount of persuasive machinery—burning bushes, hovering spirits, and the like. In any case, science is certainly no exception.

Scientists, although they are as fond of philosophers (from whom they come) of finding hard-boiled aliases for their rhetorical activities, traffic incessantly in suasion. They argue. They challenge and criticize. They appeal to form, to prestige, even to emotion. They build elaborate discursive structures out of loaded terms, topoi, and set pieces. They change each other's minds. They sway and are swayed.

Their aliases, however, have been very successful. Their acts of rhetoric when they "do science" are usually self-effacing in the extreme, covert enough that they appear to be out of the domain of rhetoric altogether. Their overt acts of rhetoric—polemics, Nobel acceptance speeches, grant proposals, and the like they disclaim as something other, something which is wholly incidental to science. Like Socrates, they differentiate "real science" from all those other things they do with words, even when they do them as scientists. Consequently, while other humanist approaches to science (notably in history and philosophy) have venerable traditions, rhetoric of science is still a relatively new and quite apologetic discipline. The apologia can go overboard. Jerry, for instance, and McGuire trade on the curious perspective that "science is not truth but a form of lie" to argue that rhetoric has a moral duty to act as "a counter-force" against the inflated prestige of science [2, 3]. But, although it is not necessary to drag science down into the covinous muck where many relegate rhetoric, the impulse to let a little air out of the word is on the right track. In particular, scientists are said to deal in Truth and Certainty and Rationality, when, in fact, they deal with the same deflated phenomena that occupy most rhetoricians, truth and certainty and rationality. Where the two differ is at the intersection of these three discursive forces, knowledge. Scientists make it. Rhetoricians, for the most part, study it, and study the making of it.

Cutting to the chase: scientists make knowledge because they are rhetors, and this paper looks at the twin paths scientist-rhetors break in pursuit of knowledge. One of these paths is dissent—the immensely productive, back-biting, barking way that scientists forge truth. The other is assent—the smoothly pervasive, communal, cooperative concert in which they arrange their truths into knowledge. These two paths give a distinctly non-Socratic meaning to the Spartan's observation: truth and discourse are fundamentally inseparable. They are Siamese twins, sharing the same vital organs. They are weight and mass, chaos and order, wave and corpuscle. You can't have one without the other. And you can't have science without both.

## 2. RHETORIC IN SCIENCE

*Please observe, gentlemen, how facts which at first seem improbable will, even on scant explanation, drop the cloak which has hidden them and stand forth in naked and simple beauty.* —Galileo

Science is profoundly rhetorical. Ignoring the many intermediate patterns, take only the two extremes that Kuhn identifies: normal science and revolutionary science. Rhetoric permeates the one and constitutes the other. Rhetoric suffuses normal science so completely that it is impossible to find even the smallest corner without suasion, argumentation, topoi—without the nuclear ingredients of rhetoric with which we manage symbols, achieve consensus, and make knowledge. Normal science is unimaginable without agreement, and agreement is unimaginable without rhetoric. At least, to rhetoricians. Scholars with ostensibly opposing positions (that agreement occurs in spite, not because, of rhetoric), such as most philosophers and most scientists, have an impoverished view of rhetoric and simply use one of its aliases to talk about consensus. As for revolutionary science, even those who deny the deeply rhetorical nature of normal science drop the aliases here and concede that periods of scientific dissent are openly rhetorical. They also usually concede that these periods are epistemically very productive.

As in all other domains, then, rhetoric has two principal reflexes in scientific argumentation, to achieve consent, and to galvanize dissent. Both reflexes have received a good deal of attention in the various humanist approaches to studying science. Traditional philosophies of science emphasize the criteria by which scientific communities reach agreement; traditional sociologies emphasize cooperative cohorts; traditional histories of science emphasize the steady accretion of knowledge. More recent philosophies of science emphasize dissonance; sociologies, the clash of incommensurate positions; histories, periods of upheaval. But it is important to realize that—as Kuhn captures in a rare eloquent phrase, *The Essential Tension*—both sides of this dialectic rule science equally.

There is agreement. Scientists cooperate. Knowledge grows. Mitchell Feigenbaum, a physicist currently embroiled in the cooperation and anarchy that characterize the emergence of chaos studies, puts it this way: today science "tells you how to take dirt and make computers from it." Yesterday it told us how to take dirt and make lenses from it that brought the stars closer and made the microorganic world grow bigger—close enough and big enough to enter our theories of the deepest mysteries, life and death. Tomorrow, if the men who play with the nastier fruits of science give us one, it will tell us how to take dirt and make eyes and hearts and consciousness from it. And this knowledge will be, as it is now, as it was in the past, the product of cooperation and consensus.

It will also be, as it is now, as it was in the past, the product of exclusion and dissensus. This is certainly true observationally: the accretion of knowledge is far from linear. Science is gangly. It takes two steps forward, one step back, another to the side. There are relatively long periods of stasis, but the equilibrium is punctuated by fierce, uncompromising, unruly debate, by rhetoric which only "enspirit[s] the already enspirited troops while further enraging the already enraged enemy." But, more crucially, this punctuation is not an unfortunate interruption in the smooth flow of a single, grandly unfurling tapestry of truth, an epistemic hiccup. It is a necessity, an engine of knowledge and truth. Fierce, unruly debate is at least as productive as calm, orderly cooperation. Often, it is much more productive. Paul Feyerabend, for instance, has built a compelling cluster of ideas about scientific knowledge-making around the notion of counter-induction—in which a theory makes truth, earns its bacon, not primarily by amassing good evidence (induction), but by amassing good counterevidence to prevailing dogma. As is his own countervailing wont, where most commentators place this cluster of ideas in the province of philosophy of science, Feyerabend says it belongs to anthropology of science [7]. But Cook and Seamon are a good deal closer to the mark in classing it with rhetoric of science especially in light of Feyerabend's own connection of counterinductivity to what he sees as the sophistic goal, *"to make the weaker case the stronger"* [8, 7].

Notice, however, that we are back to only one face of the dialectic. This is rhetoric, clearly, but it is a long way from the rhetoric of good reasons. Rather, it is a rhetoric of bad reasons—or, if there is a distinction, a rhetoric of good counter-reasons. It does not pursue consubstantiality, but dissubstantiality; it is not rhetoric of assent, but rhetoric of dissent.

Rhetoric of science, though it has not become sophisticated enough theoretically to bring Wallace or Burke or Booth to the task in any concerted manner, has made a number of very impressive sallies up both faces of rhetoric in science, the management of consensus and the productivity of dissensus. Campbell's several papers on Darwin, for instance, demonstrate how he engineered agreement by plugging into the most powerful currents in the ethos of Victorian England—chauvinistic nationalism, bootstrap capitalism, and natural theology [9]. Halloran shows much the same about Watson and Crick, for a strikingly different and much more local community [10]. Lyne and Howe scale the other face of the dialectic, exploring the epistemic fruitfulness of the phyletic gradualism-punctuated equilibria controversy in the neo-Darwinian synthesis [11]. Anderson shows the same with respect to Lavoisier, who redirected the course of science, in large part by manufacturing a dispute with phlogiston chemistry [12]. And Gross spends time on both faces, exploring first how Newton failed to affect the science of optics when he approached it trying to pick a fight with Descartes, and then how he revolutionized it several years later with a conciliatory offering of the same ideas [13].[1]

Although these papers, and others of varying quality, clearly demonstrate the capacity of rhetoricians to make important contributions to humanist studies of science, and although there has been a good deal of talk in the last decade or so about the epistemic consequences of rhetoric, most rhetoricians still have a sneaking wariness about the legitimacy of their colleagues speaking about how scientists make knowledge.[2] The reticence is understandable, since *science* is such a swollen and honored word in our culture, and *rhetoric* is such a small, ignoble one. The god-term and the weasel-word parcelled together in a construction like *rhetoric of science* still seems incongruous, even for professional rhetoricians, and leads to a fair amount of squirming at rhetoric conferences.

A more appropriate reaction to that construction, however, would be to observe that it is virtually redundant, on a par with *rhetoric of politics*. Like politics, science is so thoroughly saturated with rhetoric there is very little room for anything else. More pointedly, just as in politics, that is not a bad thing. Rhetoric is not an unfortunate social disease corrupting an otherwise robust and virtuous activity. Quite the opposite, as even extremely hard-nosed theories of science, like most varieties of positivism, tacitly admit. Positivism is, after all, a semantic theory, concerned exclusively with building words into explanatory structures and with closely monitoring meaning within those structures—two fundamentally rhetorical pursuits. It is also a theory of how to build words into arguments that produce conviction, an idealized and unreal variant of the central notion in rhetoric, persuasion.

Still, the acknowledgement is far from overt and, although more flexible and genial notions of science are currently the top sellers in philosophy, a few observations to smooth this discomfort are in order. In particular, we need to consider how thoroughly rhetoric permeates the bones of science, and how, like calcium, it strengthens those bones, giving science a spine sturdy enough to support the back-breaking work of digging up truth and building knowledge.

## 2.1 The Pervasiveness of Rhetoric in Science

*In all eternity it is impossible for me to compel a person to accept an opinion, a conviction, a belief. But one thing I can do: I can compel him to take notice.* —Soren Kierkegaard

---

[1] Gross [in *On the Shoulders of Giants*], however, does not pay sufficient attention to Newton's different public ethoi when he made his two offerings. The *Principia* came in between, giving his words on optical theory considerably more authority in 1704 than they had in 1672.

[2] See J. Zappen for a good survey of this literature [14]. Lyne and Howe p. 142, n. 2, contains a few additional references [11].

Most rhetoricians are hindus in one essential regard. Hinduism holds that everybody is a hindu, but most people are just too unenlightened to know it, and rhetoricians hold that everybody is a rhetor, but that most people are just too unenlightened to know it. Certainly some of us feel this way with regard to those people who constitute the various sciences. But even rhetoricians often fail to see that the relationship between rhetoric and science is not one of involvement, but of commitment. (To illustrate the distinction, in a bacon and egg breakfast: the chicken is involved; the pig is committed.) Science is rhetorical, leaf to root, but most scientists and many rhetoricians regard it as a dispassionate, objective activity, and, replete with upper case reverence, an eminently Rational one. They regard science as fundamentally above the reach of the petty suasions that govern life in other, less rational, spheres. But the evidence is compelling that scientists are deluded rhetors.

As a deep example of the pervasiveness of suasion in science, consider the well-known, but little discussed "experimenter effect" in research with human subjects. Even when the experiments are conducted and the data gathered by "blind" assistants (researchers who are deliberately kept ignorant of both the hypothesis being tested and the predicted outcome), there is measurable bias in the results, and the bias is uniformly in favor of the predicted outcome. Indeed, even the much vaunted "double blind" technique exhibits the experimenter effect in a significant number of investigated Cases [16]. Clearly, there is *a* marked, if subliminal, amount of persuasion at work here. The experimental designer conveys her expectations to her assistant, who conveys them to the subject, who behaves in accordance with the designer's wishes, unbeknownst to all.

Or, consider a less opaque example of this pervasiveness. Consider the impact that names have on gaining adherence for, and shaping, a position. Stephen Hawking, for instance, considers John Wheeler's name for a certain class of celestial and theoretical phenomena a "masterstroke" because *black hole* "conjures up a lot of human neuroses" and consequently focuses a great deal of attention on those phenomena, determining the concerns of scientists and their theories.[3] In Kierkegaard's terms, it compels them to take notice. Noam Chomsky's *deep structure,* though he has since repudiated it for causing "confusion. . .at the periphery of the field," was a similarly compelling masterstroke [18]. Because of the blurring between *deep* and *profound,* Chomsky's term helped garner a staggering amount of scholarly attention to his theories, as well as a fair amount of unscholarly attention. The greater part of a very vibrant decade of linguistic research hinged on competing interpretations of

---

[3] Quoted in J. Boslough [16, p. 66]. See G. Greenstein for some interesting, not to say peculiar, meditations on these neuroses [17].

the term; it is difficult to imagine a denotatively comparable but connotatively more neutral term (say, *initial phrase marker*) determining the flow of so much energy. Gell-Mann's famous borrowing of *quark* from Joyce also generated a great deal of attention, and helped shape the ethos of an entire discipline in the bargain—a discipline which now rejoices in generating terminology like *charm* and strangeness and *color* to code properties of the quark. Even more strikingly, consider how the tyrannizing image of language has determined molecular biology, where they talk of *codes* and *translations* and *semantics* as if molecules really traffic in meaning. The entire science would be unthinkable without the metaphor and the terminological baggage it brings along. Nor is this a new phenomenon. Particle physics of the mid century, for instance, is saturated with the language "of disintegration, violence, and derangement," language which was very compelling for the existential intellectuals of the period [19]. The Devonian controversy in Victorian geology saw the introduction of terms like *Silurian* (after a British tribe) and *Cambrian* (from the Roman name for Wales), reflecting the chauvinism of nineteenth century Britain.[4]

Even further back, Enlightenment scientists extended their reverence for individual men of genius to the naming of celestial objects, the Romans extended their religious concerns, the Babylonians their kinship preoccupations.

Still more clearly, consider the amount of argumentation in science. Even the desiccated remnants of these arguments that make it into the mainstream literature—where persuasion is said to be replaced by harder-edged notions, like demonstration, proof, and conviction—are rhetorical through and through. Although the rational processes behind the argument is often hidden, and the suasive intentions obscured, the style, the terminology, the construction of an exigence, even the use of citation in scientific articles, are all chosen more or less carefully (and, hence, more or less successfully) to sway the reader into the author's perspective. The format of the experimental paper, for instance, masks a great deal of the experiment it reports, while serving the latent rhetorical function of appealing to power of induction [21]. The level of readability in scientific prose also has suasive ends, and even the lowly footnote is often wielded very deftly [22, 23]. Scientific presentation "may be dressed up to resemble the policeman's deadpan testimony in the witness box; but the true analogy is with the barrister's advocacy, designed to sway the jury to a favorable verdict" [24]. Even that model of rationality, Isaac Newton, illustrates this point clearly. He worked out most of his celestial mechanics with a mathematical tool he developed for the purpose, calculus, but he presented his conclusions in the language of geometry, to put them in a more common tongue, and to borrow from the authority of Euclid and

---

[4] See J. A. Secord for an examination of, among other interesting issues, names in early Victorian geology [20].

Apollonius. He was also extremely adept at manipulating the fudge factor, and a large measure "of the *Principia's* persuasiveness was its deliberate pretense to a degree of precision quite beyond its legitimate claim" [25].

And, of course, there is no shortage of less subtle rhetoric in science. Consider this little agonistic gem by Nobel lauriate Sheldon Glashow:

> Until the string people can interpret perceived properties of the real world, they are simply not doing physics. Should they be paid by universities and be permitted to pervert impressionable students? Will young Ph.D.'s whose expertise is limited to superstring theory be employable if and when the string snaps? Are string thoughts more appropriate to departments of mathematics or even to schools of divinity than to physics departments? How many angels can dance on the head of a pin? [26].

All the stops are out here. Glashow is attacking a new theoretical school in physics with a whole range of *ad bacculums*—moral warnings, financial threats, threats of excommunication—which build toward a nose-thumbing flourish of ridicule that stretches back to Bacon's assault on the Schoolmen. Such manifest hostility is usually masked in the public face of science—though there are attested examples of public obscenity and even violence—but the fact that it occasionally breaks out so ferociously is clear evidence that there is always dissonance lying dormant in the sciences.

These examples could be multiplied at will: rhetoric pervades science like a fart in a confessional. At least, the common perception is that rhetoric is as disruptive to the quiet, pristine halls of science and as malodorous as such excretions. Certainly Kuhn, the first man to draw attention to this pervasiveness—with any degree of rhetorical force, was treated as though he had fouled the monastery. He was also treated as something of a traitor, a confessor of secrets, the one who let it out. And there turned out to be more than a grain of reality to these fears. Before Kuhn (and Lakatos, and Feyerabend) had any impact on the popular conception of scientists, they were seen as detached, ascetic authorities, untroubled by the egos, passions, and paranoia that characterized the rest of the human barnyard. At best, they were the priests of a new age, at worst, nerds. Now they are regularly portrayed as "howling, scrapping alley cats" in books like *Bones of Contention, Nobel Dreams: Power, Deceit and the Ultimate Experiment,* and *Betrayers of the Truth* [27]. But if Kuhn brought the bad news, the good news is that the deeply rhetorical nature of science also accounts for its phenomenal knowledge-making success.

## 2.2 The Productivity of Rhetoric in Science

*It is only through the clash of adverse opinions that the remainder of the truth has any chance of being supplied.* —John Stuart Mill

Again, the number of events and episodes which illustrate the scientific productivity of rhetoric is, to all practical extents, limitless (and, in many cases of course, coextensive with illustrations of pervasiveness). But, changing tack a few degrees, one representative anecdote should establish the point firmly.

Richard Muller tells how he was quietly minding his own business in his office one day when Louis Alvarez burst in with "Rich, I just got a crazy paper from Raup and Sepkoski. They say that great catastrophes occur on the Earth every 26 million years, like clockwork. It's ridiculous" [28]. Alvarez had been one of the prime movers and, thanks to a rich vein of native obstreperousness, one of the prime media figures, in the catastrophic extinction story about the dinosaurs. He and a small group of collaborators had suggested that the dinosaurs went missing because a huge comet plowed into the earth, kicked up vast clouds of debris, choked off the sun, and disturbed plant production so much that the biggest and most vulnerable members of the food chain couldn't get enough to eat. This hypothesis stirred up a great deal of rancor, with paleontologists yelling that Alvarez and his team didn't have the ethos to make these arguments, since they came from other disciplines (chemistry, geology, and physics), and Alvarez sneering back that "[paleontologists] aren't very good scientists. They're more like stamp collectors" [29]. But the claim that this was only one of a regular series of extinctions, a claim coming from a pair of paleontologists no less (even though Raup had been sympathetic to the Cosmic Interuptus theory of dinosaur extinction), was too much.

Muller agreed. Periodic extinctions *"did* sound absurd," but he was cajoled into playing devil's advocate. Alvarez had written a letter to Raup and Sepkoski, pointing out the errors of their ways, but he wanted Muller to look it over first, with *look it over* as a euphemism for "study the paper, the letter, and all pertinent data, then work your damndest to find holes in all three." Alvarez had an especially eristic view of how science should proceed, reflecting Mill's maxim about the fruitfulness of clashing opinions, and Muller had been his most promising apprentice at Berkeley; he went away fully expecting Muller to find some strong counter-arguments to his case against Raup and Sepkoski.

When Alvarez came back a few days later, after Muller had thought carefully about the issues, Muller took his role reluctantly, "like a lawyer, interested in proving my client innocent, even though I wasn't totally convinced myself." Alvarez quickly mounted a belligerent offence, pulling authority, calling names, and refusing to grant any merit at all to Muller's strongest and most obvious counter argument: that, assuming some arbitrary mechanism which could slam an asteroid into the earth every 26 million years, the Raup and Sepkoski data held up. Alvarez was completely obtuse to the point, repeatedly demanding "What is your model?" Muller said that he didn't need a model. The mere possibility of such a model legitimated the data that Alvarez wanted to discount. Muller argued that the shoe rightly belonged on the other foot, that Alvarez would have to demonstrate the impossibility of such a model. They were at a particularly fierce onus probandi loggerhead.

Alvarez turned up the heat. Muller became desperate ("Why couldn't Alvarez understand what I was saying? He was my scientific hero. How could *he* be so stupid?") and decided he would win the argument on Alvarez's terms. He grasped at the first model that occurred to him:

> Suppose there is a companion star that orbits the sun. Every 26 million years it comes close to the Earth and does something, I'm not sure what, but it makes asteroids hit the earth. Maybe it brings the asteroids with it.

Alvarez lapsed into uncharacteristic silence. He had been deliberately baiting Muller to come up with a model, but only because he was holding some damaging data up his sleeve with which he could shoot down the obvious class of explanatory models; namely, that some agent external to our immediate solar neighborhood intruded periodically.[5] But, since the hypothesized companion star and its asteroids would have the same chemical signatures as materials in our solar system, Muller's model, though very outlandish, was not vulnerable to his counterattack.

The argument cooled down almost immediately and the two of them began working out the specifics of whether or not the orbit was feasible, what effects nearby gravity fields would have on the stability of the companion, how far it would have to be away from earth now, and so on. Within minutes the baiting, the name-calling, the rank-pulling, and the belligerence had evaporated:

> It looks good to me. I won't mail my letter. Alvarez's turnaround had been as abrupt as his argument had been fierce. He had switched sides so quickly that I couldn't tell whether I had won the argument or not. It was my turn to say something nice to him, but he spoke first. Let's call Raup and Sepkoski and tell them that you found a model that explains their data [28, p. 3].

---

[5] Alvarez knew of a recent finding that the geological layer associated with the extinction of the dinosaurs has an isotopic signature (a rhenium-187/rhenium-185 ratio) that matches the rest of Earth's crust, indicating that whatever caused the catastrophe which eliminated the dinosaurs was formed at the same time as our solar system, hence was very likely formed as part of our solar system. Ironically, this is a pretty weak counter-argument, since it is fairly easy to imagine a model in which some external agent disturbed asteroids or comets within the solar system which then plowed into the earth. Indeed, even if Muller's star was formed in another part of the galaxy altogether, at another time, and therefore didn't have the rhenium signature (if, say, it was passing by and was somehow trapped into a new orbit by our sun), Muller's theory would have exxactly the same practical consequences (disturbing the Oordt comet belt and hurling some of those comets into the inner solar system).

The moral of the story is that fierce, unconstrained, argumentation is productive. Muller's hypothesis has no direct empirical support yet, but it is a very hot topic in astronomy and paleontology at the moment—generating papers, conferences, and arguments of equally fevered tenor, and it has a large body of astronomers developing methods to comb the heavens, pursuing the star which Muller has dubbed *Nemesis,* and the Nobel prize that would almost certainly follow its discovery.[6]

The Alvarez-Muller story is far from unique. I chose a relatively circumscribed event to illustrate the point, and a relatively novel one to minimize the strain on your patience, but such examples can be multiplied *ad libitum.* The most immediate examples are the ones that Kuhn brought to widespread scholarly attention in 1962, revolutions: two subcommunities go at it tooth and nail over rival explanatory programs for a class of data until, at least in the more sanguine versions of Kuhn's thesis, the remainder of the truth is supplied. In less sanguine versions (usually given as straw positions in polemics against Kuhn), there is no truth at all, since debates between competing programs "fail of objectivity" and resolutions without objectivity entails only "non-rational conversions" [31]. In brief, the argument runs: truth is an objective commodity; objectivity is unavailable in Kuhn's picture of scientific change; therefore, truth is unavailable. But such arguments manufacture their difficulties with Kuhn, by confusing truth with Truth.

## 3. TRUTH AND KNOWLEDGE

*In characterizing an episode or a state as that of* knowing, *we are not giving an empirical description of that episode or state; we are placing it in the logical space of reasons, of justifying and being able to justify what one says.*—Wilfrid Sellars

---

[6] Returning to an earlier theme for a moment, notice how the name, which quickly picked up an attendant epithet, *the death star,* is one of the reasons that Muller's hypothesis is such a hot topic. Here we also have clear evidence that Muller knew exactly what he was doing when he chose the name. He says that he was inspired by Gell-Mann's *quark* to pore over *Bullfinch's Mythology* looking for a name with just the right ring of classicism and violence. D. Raup attributes some of the theory's success to its name—which he wisely borrowed for the title of his own book on the catastrophic extinction controversies, to help ensure *its* success [30]. Muller himself uses not only the name in his title, but also its Homerian epithet. Moreover, he admits that a competing hypothesis, based on a hidden tenth planet, is "even cleverer than the Nemesis theory," adding ingenuously, "Maybe it was even true, and we should be searching for Planet X rather than Nemesis. I hoped not. I thought 'Nemesis' was a better name than 'Planet X' " [28, p. 176]. One of the principal opponents apparently recognizes the power of the name, since he refused to use it at a major conference on catastrophic extinction hypotheses, referring to it only as "the putative death star" [28, p. 154].

It has been clear since at least the turn of the century that if there is such a thing as Truth, its home is not in science. A profusion of logics, geometries, and physical hypotheses in the latter part of the nineteenth century led to the realization that scientific theories were not, as had long been the blythe assumption, "exact and exclusive duplicates of pre-human archetypes buried in the structure of things, [in]to which the spark of divinity hidden in our intellect enables us to penetrate" [32, p. 40]. An instrumentalists view of science precipitated around the dominant scientist-philosophers of the period—men like Maxwell, Ostwald, Boltzmann, Duhem, Hertz, and Mach—which recognized that theories were, necessarily, symbolic representations of nature, with no more claim to exact correspondence with nature than a string of letters (say, *theory*) has to the concept it evokes. In Boltzmann's paraphrase of Hertz,

> a theory cannot be an objective thing that really agrees with nature but must rather be regarded as merely a mental picture of phenomena that is related to them in the same way in which a symbol is related to the thing symbolized. It follows that it cannot be our task to find an absolutely correct theory—all we can do is to find a picture that represents phenomena in as simple a way as possible [33].

It cannot be the scientists' task to find the Truth, but only the truth.

This instrumentalist turn in the philosophy of science had a profound and bracing, if short-lived, effect on epistemology. It gave rise to the pragmatism of Pierce, James, and Dewey—to "more flexible and genial" notions of knowledge and truth [32, p. 41]. In essence, these more genial notions are updated variants of Protagoras' position that man is the measure of all things. They take the god-term status away from *truth*, stripping away it charismatic authority, and return it to the language a leaner, more functional term. Truth becomes a relation that pieces of language have with other pieces of language—measure and mediated by man—not a metaphysical status. Truth in this view is not doled out at Genesis, to reside as a stagnant property inherent in Ideas that float in another realm: "truth *happens* to an idea. An idea *becomes* true, is *made* true by events" [32, p. 97].

It is important to notice that these instrumentalists and pragmatists were writing during the absolute apex of classical physics—a time when it seemed so certain that a prestigious professor whom Planck approached could tell him that "Physics is a branch of knowledge that is just about complete. The important discoveries, all of them, have been made" [34]. And while most philosophy quickly returned to exploring Platonic and Kantian notions about certainty (until the more flexible and genial developments of Sellars and Quine and Rorty), science became more instrumental and pragmatic, in deed if not always in word. Planck developed quanta, Einstein relativity, Bohr probabilistic models of subatomic behavior, Heisenberg the uncertainty principle—to stay only with physics. All of these developments, and a great many more, drove science further away from its allegiance to certainty, most explicitly in Heisenberg's case, and in Einstein's famous aphorism "as far as the laws of mathematics refer to reality, they are not certain; as far as they are certain, they do not refer to reality" [35]. But a lack of absolute certainty did not suggest to these men that quantum mechanics or relativity was false—Einstein was so sure of relativity's truth that he was notoriously unfazed by empirical arguments one way or the other—just that their truth was contingent [36]. James points out that a Beethoven string quartet can be accurately and exhaustively explained as the scrapings of horses' tails over dried cat bowels, but there are other truths which hold for the same event [37]. And these additional truths can augment, modulate, or replace that account, depending on the criteria relevant to the describer and his audience.

Moreover, these scientists made truth the old fashioned way. They hurled adverse opinion at one another until the remainder was supplied. At various levels of hostility, and to various degrees of return fire, Boltzmann attacked Planck, Planck attacked Einstein, Einstein attacked Bohr and Heisenberg; Bohr and Heisenberg had fierce arguments.[7] But hurling adverse opinion is only half the story; the more dramatic half, to be sure, but half all the same. The step from truth to knowledge is agreement, and all of these disputes ended in agreement. The agreement was not always between the principals of the dispute (Einstein and Bohr never did agree about quantum theory), though such resolutions are clearly possible (Bohr and Heisenberg did), but there is now very widespread consensus that quanta, relativity, probabilistic subatomic behavior, and the uncertainty principle are among the soundest, most robust, pieces of knowledge in our culture.

The fight we looked at in some detail, the one between Muller and Alvarez, serves to illustrate three things about the rhetorical dimensions of these notions—instrumentalism and pragmatism—and consequently about the rhetorical dimensions of science. First, although revolutions are the most spectacular manifestations of dissent in science, they are far from the only ones. Smashing ideas against one another, like neutrinos, is productive even at very small scales. The brief debate between Muller and Alvarez has sparked a great deal of work, which, even if the big idea at the center of their theoretical instrument fails to stand up, if Nemesis is not confirmed, will no doubt spin off a great many epistemic by-products, the way NASA gave us powdered drink and non-stick frying pans. There will be subsidiary truths. Second, like those between Bohr and Heisenberg, the quarrel ended in mutual agreement. If this agreement ripples out into the general scientific community, especially if it is consonant with other pockets of agreement, particularly the ones that concern the hermeneutics of the sky, it will

---

[7] Virtually any text on the history of physics tells us of these, and many other, confrontations. See, for instance, L. Cline, where they are discussed respectively [34, pp. 49, 120, 235-244, 201].

make new knowledge. Alvarez, for one, is certain Nemesis will be in harmony with astronomical hermeneutics, but Muller is still hedging his bets.[8]

Third, and most crucially, it illustrates the fundamental epistemic function of rhetoric, which builds the theoretical instruments of science and finds their warrants. Muller gave Alvarez a good reason to believe that Raup and Sepkoski's data could be real (not just statistically artifactual). The reason it took to please Alvarez is curious, as is his initial intransigence, but when he was finally compelled to take sympathetic notice of the data, he granted his assent. The two of them then set off immediately, building a logical space of reasons to justify asserting that Muller's model was true. This job is not over, of course, and may not be for a very long time. The essential piece of support, the essential reason, is still missing. If Muller, or someone else, can successfully assign an interpretation to the sky that fits both a broader network of reasons, woven by the historical community of astronomers, and the specific demands of his model, if he can "find Nemesis," the bulk of the job will be over, astronomers will grant their assent, and we will have a new robust piece of knowledge. (Alternatively, someone could "find Planet X" and vitiate many of the reasons to believe in Nemesis.) [9]

The central feature of this process is unquestionably rhetorical. As the sophists argued two millenia ago, and more recent scholars have tenuously echoed, the business of making knowledge would not get very far without talk [38, 39]. Even the most sacred of our truths depend on what rhetoric can do for them:

> If we think of our certainty about the Pythagorean Theorem as our confidence, based on experience with arguments on such matters, that nobody will find an objection to the premises from which we infer it, then we shall not seek to explain it by the relation of reason to triangularity. Our certainty will be a matter of conversation between persons rather than a matter of interaction with nonhuman reality [40, p. 157].

Rorty is quite explicit on this point, echoing a long line of clear thinkers, running through Berkeley back at least to Heraclitus: it is extremely difficult to root knowledge and truth directly in nonhuman reality, in the ontology of notions like triangularity. It is also, of course, extremely difficult to root it in discourse, a notoriously loose and shifting soil. But discourse has the virtue that it houses our

---

[8] See Alvarez's preface to Muller's book and Muller's remarks about Planet X in footnote six above [28].

[9] There is at least one other point that Muller's discussion raises (as does the instrumentalist discussions above): that the depiction of scientists as deluded rhetors given earlier in the paper is something of a caricature. Muller is, however, unequipped with the analytical machinery of rhetorical theory, quite aware of the discursive, argumentative, suasive aspects of his field. Raup's book on the same set of disputes reveals an even deeper awareness of the extent to which "science is basically an adversarial process," and, in general, the closer a scientist is to controversy, the more aware he is of the productive nature of dissent.

truths, shelters our knowledge. And we can come to understand them better if we examine the architecture of that dwelling than if we stumble blindly outside searching for an ineffable triangle.

I hold with many of the sophists—as well as with James and Rorty and others who uses different terms—that rhetoric makes truth as well as frames it (more particularly, that it makes it as a consequence of framing it), and this paper argues from that position in several places. But there is no need to grant this strong form of the rhetoric-as-epistemic thesis to see that knowledge would be oarless up a creek without rhetoric. Suppose that the Pythagorean Theorem is somehow True, that it is a property of something, somewhere, an Immutable Idea in the Realm of the Forms, and not solely an instrument of human minds. Such a reaction is natural. "When you discover these things," says Feynman, "you get the idea that they were somehow true before you found them. So you get the idea that somehow they existed somewhere." He also adds, however, that "there's nowhere for such things" [41, p. 208]. But let's ignore him. Let's arbitrarily deny the first of Gorgias' three epistemic-cum-ontological principles. We are still faced with two more. Let's say there is such a place. How would we discover the Truths that lie there? How would we grant Them our assent, share our warrants about Them, teach or be taught? How would we *know* without the measuring, mediating, human force of rhetoric? We wouldn't.

### 3.1 Dissent

*Whereas unanimity of opinion may befitting for a church, or for the willing followers of a tyrant, or some other kind of 'great man,' variety of opinion is a methodological necessity for the sciences.*—Paul K. Feyerabend

Rhetoric and hostility have always been close travelling companions. Plato, for all his suspect motives, hits pretty close to the mark when he lampoons a group of sophists who used to be teachers of warfare, and more recent observers have noticed that martial language defines the way we talk about argumentation [42, 43]. Returning to science, and to a specific example, consider the vitriol from Glashow above. It is an unmistakable symptom of the Max Planck Effect—the phenomenon whereby two opposing camps divide over paradigmatic issues along generational lines. The young Turks propose some conceptual reorganization, and their elders in the field repress that reorganization. Acrimony wells ups. Ridicule and political manipulation rapidly become the prime instruments of attack. The youngsters thumb their noses at the failures of the established program, and warn that anyone who tries to maintain "the old religion" will be swept away in the tide of historical change [28, p. 152]. The elders sneer at the vast array of problems the new program can't address and oppose it on the grounds that it will strand the next generation of scientists up a blind alley with no map to guide them back to the real issues of the field. In this case, Glashow defends the standard quantum mechanical view of physics that he helped to build, against the heresy of a group of theorists who are attempting to build another view of physics around undimensional curves they call *strings*. They, of course, are fighting back, making a good deal of hay out of the inconsistencies of the standard quantum program (in particular, that all attempts to reconcile it with gravity generate absurdities). Glashow, in turn, is very frank in his repression: "I do everything in my power to keep this contagious disease—I should say far more contagious than AIDS—out of Harvard" [41, p. 157]. *Ad hominems* are the order of the day; the Turks suggest their elders don't have the intellectual chops to follow string theory because it depends on "real mathematics, not ersatz mathematics," and the Old Guard snaps back that "the mathematics is far too difficult for [the string theorists], and they don't draw their conclusions with any rigor. So they just guess" [41, pp. 179, 194].

The Max Planck Effect is just one aspect of clashing opinions in science, which very frequently involves animosity, which always involves dissent, and which is one of the prime vehicles whereby scientists make knowledge. *Opinion*, though, is rarely the word that antagonists in such a dispute choose to characterize their positions; in fact, the preferred term is a word usually taken to be its virtual antonym, *truth*. Each side is convinced that it has the program which, in the terms it chooses to frame the debate, has the brightest spark of truth to it. Each side is Certain. Yet Certainty is only supposed to obtain, as Descartes told us, when there is no room for doubt, and it is an important maxim of epistemic investigations of science that there is always room for doubt. No theory is ever fully determined by the evidence. What the underdetermination thesis should lead to, in a perfectly rational world, is a high degree of skepticism, with each scientist placing bets on the approach that seems most fruitful at the moment. And, in effect, that's pretty much what goes on as a general trend in science. But in times of conflict, when disparate paradigms clash head on, scientists on either side of the schism suspend their skepticism for the fundamental principles of their own paradigm, and magnify it for the competing paradigm. From largely the same, necessarily impoverished, body of evidence, each side is certain about its own claims to truth.

Robert Thouless noticed this phenomenon much earlier in the century, and framed it, as was the positivist wont of social scientists at the time, as a law of social psychology:

> When, in a group of persons, there are influences acting both in the direction of acceptance and rejection of a belief, the result is not to make the majority adopt the belief with a low degree of conviction, but to make some hold the belief with a high degree of conviction while others reject it also with a high degree of conviction [44, p. 24].

Thouless called his maxim the *principle of certainty*, because it notices that in precisely those instances where doubt should intercede, when there are good reasons on both sides of a question, people tend to be most certain about their beliefs. Edward Witten, for instance, one of the "Princeton String Quartet," is so convinced of the truth of string theory that he calls it "a piece of twenty-first century physics that fell by chance into the twentieth century," while Glashow is equally certain that Witten's dates are vastly off target, suggesting rather that string theory is a sterile piece of thirteenth century scholasticism that has infected twentieth century physics [45].

Thouless attributes this impulse to cognitive dissonance—"doubt and skepticism are for most people unusual and, I believe, generally unstable attitudes of mind"—and he connects the desire to avoid dissonance quite explicitly to the sorts of epistemic shirts now familiar in the study of science:

> It may be that the operation of this tendency is a considerable part of the explanation of sudden intellectual conversions, in which a new opinion comes into the mind with strong conviction as a result of the spontaneous tendency of the mind to pass from the unstable and painful condition of doubt to the stable and tensionless one of certain conviction [44, p. 29].

One of the interesting aspects of Thouless's position here is that it is shared by James, who says that when you have knowledge, "epistemologically, you are in stable equilibrium" [46, p. 97]. Agitation is valuable, then, because it compels people to pursue the stability of sure knowledge. But James differs in two essential points, both of which connect him with sophistic lines of thought, and both of which make very clear the epistemic exigence for rhetoric of science (and, incidentally, partially mark the boundaries between psychology of science and rhetoric of science). First, whatever the psychological drivers which career people toward knowledge, truth is more the trajectory than the destination, more process than product. The meeting of belief and reality is not a matter of "truth or consequences" for James, but of "truths are consequences." "True ideas" for him "are those we can assimilate, validate, corroborate and verify" [46, p. 41]. At its most idle, truth is a stored charge, an action potential. A belief must have, ever ready, a battery of corroborative and verificational arguments, or it is not true.

Second, and more important, James holds that it is right to be certain in such instances. Again, the argument involves a case deflation; this time from *Rationality* to *rationality*. If the stakes are high, it is rational to have conviction on partial evidence. If one only commits oneself to a position when it is Rational to do so, when the proposition expressing it is True, when Certainty obtains, then one misses out on the world. Since these grand upper case concepts are never attainable, a man waiting for Certainty is like a man who hesitates "indefinitely to ask a certain woman to marry him because he was not perfectly sure that she would prove an angel after he brought her home. Would he not cut himself off from that particular angel-possibility as decisively as if he went and married someone else?" [37, p. 26]

When the stakes are high, as they are in paradigm clashes and marriages, skepticism can be debilitating. Witten is right to believe that string theory is a gift from the twenty-first century, because if he didn't, if his collaborators didn't, if other scientists in parallel situations didn't, as Kepler believed his insights were the gifts of God, and Darwin the gifts of Induction, there would be very little progress. What progress there was would be piecemeal, following the simple, linear function, accretionary model that Kuhn displaced. There would be no conceptual, shake-ups, no passion. Astrophysicists would be trying still to get epicycles more and more predictively accurate. By exactly the same token, Glashow is right to retrench so deeply against string theory. If there wasn't a huge amount of inertia to overcome when new perspectives surface, our communal understanding of the cosmos would rise and fall with the hemlines, a situation just as epistemically debilitating as skepticism. Within the argument fields of science, as Popper tells us, "the dogmatic scientist has an important role to play. If we give in to criticism too easily, we shall never find where the real power of our theories lies" [47]. Popper's student, Feyerabend, refracts dogmatism into his principle of tenacity, which, in concert with his proliferation principle, concisely describes epistemic progress as "the active interplay of various tenaciously held views" [48].

When these tenaciously held views clash head on over foundational issues, we call the events *revolutions,* after Kuhn's generalization of a long established term for the great period of intellectual upheaval that spun medieval epistemic assumptions into their modem shape, *the* scientific revolution. The word is apt, not the least because it makes clear that what happened to western intellectual culture in that period is played out regularly on a smaller scale in individual scientific fields. But scientific change is too heterogeneous, and dissent is too widespread, for the naive application of Kuhn's normal-science-to-revolution-back-to-normal-science paradigm. There are periods in every discipline when dissent forms along broad lines and voices get louder because they're shouting in unison—when disagreements spill out into other fields and into the newspapers—but there is always at least a slow boil, a background growl of dissonance, in every field productive enough to be awarded the honorific label of *science.* Even with those periods in the development of a field that fit the concept of normal science very closely, there are innumerable disagreements about individual problems and solutions, about the character and meaning of data, about the status of anomalies, about a whole welter of empirical, methodological, and philosophical issues not to mention the more overt power struggles over institutional organs, publication, tenure, and all the social trappings of the field.

The crucial point for our purposes is that all of this dissent is immensely productive. Science is a knowledge-generating activity, and knowledge is too precious a commodity to lay fallow for long. Rhetoric—reasoned exchange, as well as bickering, back-stabbing, barking disputation—is the life's blood of science. The point is extremely easy to establish. It is illustrated by virtually any major dispute in any science. There is usually a clear team of winners and a clear team of losers in any controversy, but if you look closely at the winning position, it almost always includes elements introduced into the dispute by the losers, elements that the winners introduced expressly to meet challenges from the losers, and elements which are simply by-products of the scuffle. That is, much of what constitutes the triumphant program, the new packet of knowledge, was either (1) appropriated from the defeated program, (2) introduced to plug holes shot in it by the gunslingers from the losing side, or (3) generated in response to issues that arose spontaneously, in the heat of battle. These elements are difficult to extract directly from the statements of the antagonists, because the winners are usually too bitter to acknowledge contributions from their enemies, and the losers are usually too bitter to discriminate. The winners (who usually write the histories) come to view their victory as an intellectual triumph over error, deceit, and pig-headedness. They are loath to see, let alone acknowledge, the portions of the other program they have swallowed. The losers, as their work and its defining context become marginalized, come to view the whole process in political, rather than intellectual, terms. They usually degenerate into crankiness, and claim the victorious program, where it is not flagrantly wrong, has stolen their truths. The resulting polemical stew follows the recipe that Keller gives for propaganda, mixing clear truth with "half truth, limited truth, truth out of context" [49]. But careful study can disentangle the components of the successful program, uncovering the network of pressures and counter-pressures which shaped it. Almost every rhetorical case study of a scientific dispute traces the development of the final positions in a dispute, revealing the sources and contexts of its constituents, but, by a vast margin, the most fine-grained study of this sort is Rudwick's map of the thrust and parry that constituted the Devonian controversy in nineteenth-century geology. Indeed, some of Rudwick's visual depictions go into so much detail, tracing the trajectories of individuals and positions in the dispute, that they resemble circuit board schematics [24, pp. 412-413].

On a more general scale, it is often the case that periods of broad intellectual turmoil are extremely productive. Certainly for the paradigmatic case of this phenomenon, the period of epistemic wrangling so productive that it warrants a definite article, *the* scientific revolution, the correlation is indisputable. The scientific achievements of the sixteenth and seventeenth centuries are best seen as symptoms, rather than as causes of a steamy agonistic climate that included a gamut of disputes personal and social, cultural and religious, political and economic [50]. And many of these controversies enter the scientific literature as themes and leitmotifs. Galileo, for instance, wrote in Italian, to enlist the clamoring mercantile classes, and he salted his arguments with analogies to the world of commerce [51]. He also made sure to appeal to the self-importance of that upwardly mobile group by contrasting their perspicacity to "the shallow minds of the common people"

[52, p. 207]. Most of the Copernican-Galilean champions were also lay people, though some were Protestant clergy, and most of their opponents were Catholic. Throughout Europe, Copernicanism came to "be construed as antipapal and hostile to the power of the Catholic clergy," and rode the crest of Protestant reformation [50, p. 25]. A classic text on the scientific revolution (and an important antecedent for Kuhn and Feyerabend) comments frequently on the epistemically "healthy friction" of the period, and glowingly commends agitators like Marin Mersenne, "a man who provoked enquiries, collected results, set one scientist against another, and incited his colleagues to controversy" [53, p. 83].

Or consider our own century. Whatever the relation to the overwhelming mood of pathological testiness that has produced two major wars, countless minor wars (if the phrase is not an oxymoron), and the recurrent clatter of jackboots, that has seen superpowers regularly explode entire islands in order to impress one another with weapons teetering on the brink of apocalypse, that daily pits state terrorism against populist terrorism—whatever the relation to this overall storm of violence, repression, and death, the twentieth century has also seen a moil of intellectual battles and an astonishing amount of new knowledge: the theory of relativity, quantum mechanics, plate tectonics, molecular biology, cognitive psychology, and array of less spectacular epistemic achievements, were all born of conflict and had to struggle for survival. All fit Kuhn's *revolution* like the premise of received violence apparently fits arguments toward the conclusion of returned violence.

The knowledge produced by these revolutions comes right from the heart of the tumult. Take again the very local example of Muller's argument with Alvarez. It is a synecdoche of *the* scientific revolution, and of most scientific revolutions. Alvarez, here playing the epistemic status quo, bullies, ridicules, pulls rank, withholds data, and dismisses counter-arguments without a hearing. He is an intolerant authoritarian, squelching dissent. Muller, holding the new position, pleads, cajoles, grasps at straws, and probably employs a number of strategies not quite so mild, which didn't stay in his memory or didn't make it into his book. He is an intolerant revolutionary, pressing for assent. Together, they forge a compelling new theory which is now in search of the empirical support needed to achieve the rank of truth. The epistemic status quo also brought Galileo before the Inquisition and threatened to burn Kepler's mother as a witch. Galileo, for his part, employed a wide range of "psychological tricks," and Kepler tried so many tacks that Holton calls his turbid style a "mirror [of] the many-sided struggle attending the rise of modern science" [7, p. 81; 19, p. 54].

Science was born of strife. Copernicus had to fight for his theories. Kepler had to fight. Galileo had to fight. And none of these fights was the fight of clear and shining Truth against the dark repression of church bureaucrats that popular mythology portrays. The scholastic world picture was full and rich. It was a largely consistent, coherent, explanatory model of the planets, of motion, of what happened when someone dropped cannon balls from the tower of Pisa. The new models (and there was an abundance of new models) quite rightly had to prove their merit in a clash with the epistemic status quo. Together, these men, their advocates and their opponents, and the fierce dialectic they all propelled, laid the cornerstones of science and the modern world.

## 3.2 Assent

*A man is necessarily talking error unless his words can claim membership in a collective body of thought.*—Kenneth Burke

For all the barking disputations of science, the product is communal assent, knowledge. The point of every argument is to gain adherence, which is to say, adherents: to increase the membership of a body of thought, to pursue assent. The first of these points is the stated exigence of one of the most lauded recent books in the history of science—that the foundation of science is the rhetoric of assent. The book, which has already made a few appearances in this paper, is Martin Rudwick's *The Great Devonian Controversy* (which Stephen Jay Gould, for instance, has hailed as potentially "one of our century's key documents in under-standing science") [54, 55]. It is a laborious case study of a forgotten geological dispute in the nineteenth century, organized around the notion that, overwhelm-ingly, the most characteristic form of scientific dispute is exactly like the Devonian controversy—a dispute which ended in a satisfactory resolution for all the principals and a new piece of knowledge. Such disputes then quickly drop out of view, and only their residue, the new piece of knowledge, remains. Rudwick doesn't use Booth's phrase, of course, but he very deliberately sets his argument against the loose philosophic confederation that Laudan calls the dissensus theorists—people like Kuhn and Feyerabend, who hold that full resolution is unobtainable for nontrivial scientific disputes, that opposing paradigms are in-commensurate [56]. Paradigm clashes, in the view of this school, can be dis-patched by political maneuvering, by superior propaganda, or simply by waiting for the Max Planck Effect to run its course, when the new program's opponents "die off, and a new generation grows up that is familiar with" [57]. But they cannot be settled in any meaningful way. Rudwick uses this school's terms for his demonstration (which, not coincidentally, are also our terms), terms like per-suasion and influence, even the nonpejorative use of rhetoric. But where Kuhn has a note of despair that a paradigm dispute is "not the sort of battle that can be resolved by proofs," and Feyerabend glories in the irrationality that necessitates rhetoric, Rudwick regards rhetoric as not only a valuable ingredient in the scien-tific ragout, but fundamental to its rationality [58, 7, p. 189]. Science for Rudwick is the human activity that takes an interpretation of the physical world and makes knowledge with it, as the interpretation is "forged into new shapes with new meanings, on the anvil of heated argumentative debate" [24, p. 455].

The crucial notion here is what Rudwick calls the "social and cognitive topography," but which we can discuss with terms like context, and frame, and circumference [24, p. 421, 59]. Rudwick's case study is essentially the close analysis of how conflicting interpretations of data pressed upon one another within the confines of the 1830s geological frame, until they fused into an interpretation that could fit the frame without too many perturbations. These interpretations would have been incoherent without that frame (or another one with similar implications). They could have neither conflicted in the first place, nor fused in the second, without the meaning lent to them by the collective body of thought that constituted stratigraphical geology in the period.

Despite Rudwick's implicit opposition to Feyerabend, the two are in complete agreement on this point, though Feyerabend makes it a little more dramatically. He argues that Michelson's set of ether drift experiments are in fact two sets of experiments—at the least, two [48, p. 152]. In one context, classical mechanics, they were a failure. They failed to generate empirical evidence of drifting ether, and to that extent, they were relevant; they undermined (but not disconfirmed) the tenets of the framework that predicted luminiferous ether. (When Michelson was awarded his Nobel prize in the distinctly classical milieu of 1907 no mention was made of this work.) In another context, special relativity, they were a success, obviating the need for an absolutely stationary space and dovetailing smoothly with Einstein's alternative framework. The point is very straightforward: "experimental evidence does not consist of facts pure and simple, but of facts analyzed, modelled, and manufactured according to some theory;" that is, of facts circumscribed by some collective body of thought [33, p. 61].

However straightforward the point is, absolutist theories of science find it repugnant. Notice, though, that abolutists wouldn't have any trouble if Feyerabend's comments exclusively concerned words. We all recognize that duck is a noun in one context, a verb in another ("Fred gave Betty a duck" and "He was lucky to duck Wilma"), and we recognize that rake is two nouns, one which describes a dissolute cad, perhaps Fred, and one which names a yard implement. Uncontroversially, this multiplicity is also true of technical terms. Democritus and Lucretius' atom, for instance, is not Rutherford and Bohr's atom. Democritus's atom, by definition, codes absolutely the smallest possible bit of matter, something which is absolutely indivisible (rather like today's quark); Rutherford's atom codes an object with constituents. For that matter, Rutherford's atom is not Bohr's, and Bohr's atom of 1913 is not Bohr's atom of 1927.

Nor do we have any trouble with sentences. We know that the string of words "Nirmala saw the man from the library" is two sentences, two propositions, depending on whether Nirmala is looking out the library window at a man, or is walking down the street and spots a man she knows to work at the library. Put in the most elemental terms with which we discuss the semantics of these complex phenomena, sentences, one interpretation can be true while the other is false, depending on the frame of reference. Again, this is uncontroversially true of technical sentences. "Parallel lines never meet" is true in some geometries, false in others. Crick's central dogma—"Information flows one way and one way only, DNA to RNA"—is true under some interpretations, false under others.

But many of us balk when something as reified as a scientific experiment is said to be ambiguous, to have more than one meaning, to be more than one phenomenon. This is simple prejudice. Experiments are much more complex arrangements of words and readings and relationships than sentences.

Rudwick's principal point is exactly this one, that the Devonian interpretation of geological strata is crucially dependent on the context of its creation; namely, the controversy that raged in and around the Geological Society of London from 1832 to 1842. The principals of that controversy ranged bitterly. They lied. They cheated. They performed anatomically unethical maneuvers, going behind each other's backs and over each other's heads. They also debated, negotiated, converged, and eventually agreed on a specific interpretation. In the course of this activity, the data changed incessantly. They had different meanings, at different points to different people, as the arguments which defined them changed. The core data set warranted different configurations of assent at different points in the controversy. Meaning, and therefore truth, modulated.

When the assent of the principals converged on one interpretation, they had made a new piece of knowledge, and now that the squabble has long since faded into history (or had, before Rudwick resuscitated it), that fossilized knowledge plays its part in the current context of geological assent.

Rudwick's wider point is more problematic. He argues that his case study is "a *characteristic* piece of scientific debate," which it surely is, but as his italic histrionics indicate, he also argues that most other case studies "are *uncharacteristic*" [24, p. xxii]. Science, he says, is full of innumerable disputes, the vast majority of which resolve into a solution that most of the principals can grant their warranted assent. The dispute then rapidly disappears into history, leaving only the remainder of the clashed opinions, the truth, as another brick in the epistemic wall, of science. He adds that humanist approaches, however, usually ignore this vast majority and focus on only the more flamboyant episodes, the ones which don't resolve smoothly, a focus which gives rise to theories of dissent, mutually hostile cohorts, and incommensurability.

His remarks hold only for *recent* humanist approaches to science, of course, but even if they held more widely, there is no reason to believe that smoothly resolved debates are more characteristic of science than any other form of debate. Certainly there are plenty of bricks of fossilized knowledge in every science, and many of them came from debates long forgotten, but that is hardly evidence that the debates all generated single solutions that all the principals found satisfactory. In the Devonian controversy, the resolution came about when the "victor" finally modified his interpretation to fall closely enough to the "loser's" position that the compromise was agreeable. They settled on a price. But if the loser had remained pig-headed to the end, and the majority of the community still found the compromise position congenial, Rudwick would surely hold (as he should) that knowledge and truth had still been served in the encounter. Or, if the loser and a significant party of backers from the general community had all remained intransigent, but had died off or were otherwise marginalized, and the victor's solution slowly took hold in the community, Rudwick would again surely hold (as he should) that knowledge resulted from the clash. When the community has sufficient warrant to reach consensus, there is knowledge.

I am not arguing that Rudwick is wrong. On the contrary, he is right. The rhetoric of science, because it is epistemic, is a rhetoric of assent. But he is right in the limited case. Assent does not come in just one characteristic flavor. More generally, Rudwick is only interested in half the picture, and is antagonistic to the other half, the rhetoric of dissent and revolution. That, he discounts as anomalous.

It should be clear by now that the rhetoric of dissent is far from anomalous, but what is more interesting for our purposes is the point Thouless raises that it is highly unstable and either rapidly evolves into arguments which seek broader adherence, or it just dies out. As Burke's observation about collective bodies of thought indicates, saying no to one body implies saying yes to another. Rejecting one circumference implies accepting another. This is perhaps one reason that scientific revolutionaries who kill off their fathers also resurrect their grandfathers; it is extremely difficult to draw up a new program *ab novo*. Moreover, the rhetoric of dissent has an especially galvanizing effect on a revolutionary cohort, which often has a level of internal assent that borders on religious devotion. Although history presents us with a pageant of great men—Copernicus, Galileo, Newton, Darwin—most intellectual revolutions can be traced to a small fervent group, banded together in the name of truth and light to battle the prevailing error of the day. The men in the pageant were certainly great, but three additional points need to be stressed. The first has already been tabled: as great as they were, they were not without their collaborators and disciples. Indeed, they had collaborators and disciples *because* of their greatness. Copernicus had Galileo, of course, and both of them had a wealth of supporters. Newton had the Royal Society virtually in his waistcoat pocket. Darwin had a brigade, headed by Huxley. Einstein had Eddington. Bohr had Heisenberg and Pauli and the Copenhagen Institute.

The second point has also made an appearance, that the sense of being privy to the truth in the face of indifference or hostility is a powerful determinant of group ethos, inducing members of the group to say yes more loudly and widely, to go forth and convert the heathens. Einstein and Schrodinger, and others, fought the Copenhagen interpretation fervently, often resorting to blatantly metaphysical appeals, like Einstein's "God does not play dice with the universe." Max Delbruck's objective in his evangelical Cold Spring harbor course in molecular biology is widely recalled as "frankly missionary: to spread the new gospel among physicists and chemists," Robert May recalls his early chaos theory call-to-arms paper for *Nature* as "messianic" [60; 5, p. 80]. Most of the early chaos papers, in fact, "sounded evangelical, from their preambles to their perorations" [5, p. 39].

The third point, however, has not yet been raised, and it is the most important aspect of winning adherents: regardless of conviction and oratorical prowess, if the missionaries don't have a positive program to offer when they denounce the old religion, there is little hope for widespread assent. Indeed, there is only a very weak form of assent called for in an exclusively negative rhetoric—a consensus of dissent, a communal agreement that something is wrong, without a clear idea of how to put it right. Einstein and Schrodinger, as passionate, eloquent, and sharply reasoned as their assaults on probabilistic models were, had no remotely comparable program to offer if Bohr's work was overturned. By contrast, Delbruck did. May has. Molecular biology rapidly became the framework that most scientists said yes to, and chaos studies are burgeoning. Scientists need something to do, and as James points out with his angel analogy, inactivity is too steep a price to pay for the privilege of rejecting error. To be successful, a new program has to have a wide enough circumference to embrace recruits, and set them to work right away. Aaron Novick recalls precisely that sense of expansiveness in Cold Spring Harbor, 1947:

> In that three-week course we were given a set of clear definitions, a set of experimental techniques and the spirit of trying to clarify and understand. It seemed to us that Delbrack had created, almost single-handedly, an area in which we could work [61].

Heinz-Otto Peitgen, one of the earliest chaos converts, describes the situation in more general terms, saying that in an established program the problems that are left to work on, the ones that don't already have consensually satisfactory solutions, have usually been passed over for a reason. They are hard. Moreover, the background required just to frame the solution is often staggering. But in a new program, like chaos theory, there are less demanding opportunities: "you can start thinking today, and if you are a good scientist you might come up with interesting solutions in a few days or a week or a month" [5, p. 230].

This, of course, is precisely why the word *evangelical* comes up so often in accounts of successful epistemic shifts. The rhetor's purpose is, in Socrates' terms, to win souls, to convert the heathens and the innocents [62]. And to win souls, rhetoric must offer consubstantiality; it must map out a place where the auditor can join the rhetor in some mutual enterprise, where they can both say yes together, not just no.

## 3.4 Conclusion

*A rhetoric which conceives of "truth" as a transcendent entity and requires a perfect knowledge of the soul as a condition for its successful transmission automatically rules itself out as an instrument for doing the practical work of the world.* —Douglas Ehninger

The structural equation woven through this paper looks a little too neat. In one line of the argument we have the productivity of rhetoric in science, the breeding of dissent, and the forging of truth. In the other line, we have the pervasiveness of rhetoric in science, the achievement of assent, and the construction of knowledge. And, of course, it is too neat. Falling back on James once again, he says about another, equally tidy, argument, "it seems *a priori* improbable that the truth should be so nicely adjusted to our needs and powers as that. In the great boarding-house of nature, the cakes and the butter and the syrup seldom come out so even and leave the plates so clean" [37, p. 22]. The productivity and pervasiveness of rhetoric in science are inextricably entangled. Dissent and assent are relative, in the most profound sense of that term. Nor are truth and knowledge easily disengaged; one is the efficacy of a belief, and the other is the consent we award a belief when our warrants convince us it is efficacious. But the two threads of the discussion, however interspliced, are two threads for all that.

As long as we have James back for a moment, in the same essay where he worries about the flapjacks of nature, he offers some help here. He identifies two deep psychological drivers which define these threads. There are two reactions we have, he says, whenever we confront the world looking for knowledge: Believe truth! and Shun error! [37, p. 17]. We want to be right, but we are nervous of being dupes. You want to agree with me, but what if I'm wrong? I want to agree with James, but what if he's wrong? We yearn for truth and fear error. Rhetoric, with its twin prongs, assent and dissent, is the discursive manifestation of both impulses.

Both prongs are associated with epistemologies of very respectable lineage. The first descends from Aristotle, had its fiercest advocate in Bacon, and is now making something of a comeback as a new inductivism, which Hacking calls "a Back-to-Bacon movement" [63]. The second descends from the skeptics and sophists, had its starkest formation in Gorgias' three epistemic principles, and has seen some latter day success under the banner of Popper's falsificationism and its various radical extensions (notably with Lakatos and Feyerabend). Feyerabend, in particular, makes the antithetical nature of these epistemologies clear, with the label he chooses for his side of the divide, *counterinduction*.

Science, the primary cultural expression of our deep epistemic urges, tries very hard to follow these two paths equally. It fails often, of course, but it also succeeds, far more often and far more spectacularly than any other cultural organ.

One of the reasons that rhetoric has had such bad press for so long, and that it is regularly portrayed as the evil opposite of science (and its parent, philosophy) is that Plato successfully pinned the scarlet disquisitional letter to it, and disquisition is relentlessly faithful to only one of James's deep rivers, Believe truth! The exclamatory formulation of this impulse is especially appropriate to disquisition, since its principal exponents are preachers, politicians, and salesmen—people whose sole objective is to exhort their own brand of the truth, and eliminate enquiry. By contrast, science has always understood—even if many individual scientists have not—that no matter how true an idea "may be, if it is not fully, frequently, and fearlessly discussed, it will be held as a dead dogma, not a living truth" [64]. Dogma has its function, as Popper, Feyerabend, and others have argued, but science has remained epistemically quite responsible because the rhetorical antidote of dissent, revolution, and renewal have always had equal play.

It has been a dogma of humanist studies of science since the 1950s that there is no such thing as *the* scientific method, but perhaps there is. Rhetoric.

## REFERENCES

1. L. F. Bitzer, The Rhetorical Situation, *Philosophy and Rhetoric, 14*, 1968; see also Functional Communication: E. E. White (ed.), *Rhetoric in Transition: Studies in the Nature and Uses of Rhetoric*, The Pennsylvania State University Press, University Park, PA, 1980.
2. M. McGuire, The Ethics of Rhetoric, *Southern States Speech Journal, 45*, 1980.
3. E. C. Jerry, Rhetoric as Epistemic: Implications of a Theoretical Position, in *Visions of Rhetoric: History, Theory, and Criticism*, C. W. Kneupper (ed.), Rhetoric Society of America, Arlington, Virginia, 1987.
4. T. Kuhn, *The Essential Tension*, University of Chicago Press, Chicago, Illinois, 1977.
5. J. Gleick, *Chaos: Making a New Science*, Viking, New York, 1984.
6. W. C. Booth, The Scope of Rhetoric Today, in *The Prospect of Rhetoric*, L. F. Bitzer and E. Black (eds.), Prentice-Hall, Englewood Cliffs, New Jersey, 1971.
7. P. Feyerabend, *Against Method*, Verso, London, 1978.
8. T. Cook and R. Seamon, Ein Feyerabendteur, *Pre/Text, 1*, 1980.
9. J. A. Campbell, Darwin and The Origin of the Species, *Quarterly Journal of Speech, 37*, pp. 1-14, 1970; The Polemical Mr. Darwin, *Quarterly Journal of Speech, 61*, pp. 375-390, 1975; Scientific Revolution and the Grammar of Culture, *Quarterly Journal of Speech, 72*, pp. 351-376, 1986. See also B. Warnick, A Rhetorical Analysis of an Episteme Shift, *Southern States Speech Journal, 49*, pp. 26-42, 1983.
10. S. M. Halloran, The Birth of Molecular Biology, *Rhetoric Review, 3*, pp. 70-83, 1984.
11. J. Lyne and H. F. Howe, Punctuated Equilibria, *Quarterly Journal of Speech, 72*, pp. 132-147, 1986.
12. W. Anderson, *Between the Library and the Laboratory: The Language of Chemistry in Eighteenth Century France*, Johns Hopkins Press, Baltimore, Maryland, 1984.
13. A. G. Gross, On the Shoulders of Giants: Seventeenth-Century Optics as an Argument Field, *Quarterly Journal of Speech, 74*, pp. 1-17, 1988.

14. J. Zappen, Historical Studies in the Rhetoric of Science and Technology, *The Technical Writing Teacher, 14*, pp. 285-298, 1987.
15. R. Rosenthal, *Experimenter Effects in Behavioral Research*, Irvington, New York, 1976 and How Often Are Our Numbers Wrong? *American Psychologist, 33*, 1005-1008, 1978.
16. J. Boslough, *Stephen Hawking's Universe*, Quill, New York, 1985.
17. G. Greenstein, *Frozen Star*, New American Library, New York, 1984.
18. N. Chomsky, *Language and Responsibility*, Pantheon, New York, 1979.
19. G. Holton, *Thematic Origins of Scientific Thought*, Harvard University Press, Cambridge, Massachusetts, 1988.
20. J. A. Secord, King of Siluia: Roderick Murchison and the Imperial Theme in Nineteenth Century Geology, *Victorian Studies, 25*, pp. 413-442, 1982.
21. A. G. Gross, The Form of the Experimental Paper: A Realization of the Myth of Induction, *The Journal of Technical Writing and Communication, 15*, pp. 15-26, 1984.
22. J. Hartley, M. Trueman, and A. J. Meadows, Readability and Prestige in Scientific Journals, *Journal of Information Science, 14*, pp. 67-75, 1988.
23. A. de Grazia, The Scientific Reception System, A. de Grazia (ed.), *The Velikovsky Affair*, University Books, New Hyde Park, 1966.
24. M. J. S. Rudwick, *The Great Devonian Controversy*, University of Chicago Press, Chicago, Illinois, 1985.
25. R. S. Westfall, Newton and the Fudge Factor, *Science, 179*, pp. 731-732, February 23, 1979.
26. S. Glashow, Tangled Up in Superstring: Some Thoughts on the Predicament Science Is In, *The Sciences, 27*, May/June 1985.
27. R. Kanigel and G. Cowley, The Seamy Side of Science, *The Sciences, 28*, July/August 1988.
28. R. Muller, *Nemesis, the Death Star: The Story of a Scientific Revolution*, Weidenfeld & Nicolson, New York, 1988.
29. M. Browne, The Debate Over Dinosaurs Takes an Unusually Rancorous Turn, *The New York Times*, C4, January 19, 1988.
30. D. Raup, *The Nemesis Affair*, W. W. Norton, New York, 1987.
31. I. Scheffler, *Science and Subjectivity*, Bobbs-Merrill, Indianapolis, Indiana, 1967.
32. W. James, *The Meaning of Truth: A Sequel to "Pragmatism,"* Harvard University Press, Cambridge, Massachusetts, (1909) 1975.
33. P. K. Feyerabend, *Realism, Rationalism & Scientific Method*, Cambridge University Press, Cambridge, United Kingdom, 1981.
34. B. L. Cline, *The Men Who Made a New Physics*, University of Chicago Press, Chicago, Illinois, (1965) 1983.
35. A. Einstein, *Sidelights on Relativity*, G. B. Jeffrey and W. Perret (trans.), Dover Books, New York, (1922) 1983.
36. I. Rosenthal-Schneider, *Reality and Scientific Truth*, T. Braun (ed.), Wayne State University Press, Detroit, Michigan, 1983.
37. W. James, *The Will to Believe and Other Essays in Popular Philosophy*, Cambridge University Press, Cambridge, United Kingdom, 1896.

38. R. L Enos, The Epistemology of Gorgias' Rhetoric: A Re-Examination, *Southern Speech Communication Journal, 42*, pp. 35-51, 1976.
39. R. Scott, On Viewing Rhetoric as Epistemic, *Central States Speech Journal, 18*, pp. 9-17, 1967 and On Viewing Rhetoric as Epistemic: Ten Years Later, *Central States Speech Journal, 27*, pp. 258-266, 1976.
40. R. Rorty, *Philosophy and the Mirror of Nature*, Princeton University Press, Princeton, New Jersey, 1979.
41. P. C. W. Davies and J. Brown, *Superstrings: A Theory of Everything?* Cambridge University Press, London, 1988.
42. Plato, *Euthydemus*, R. Kent Sprague (trans.), Bobbs-Merrill, Indianapolis, Indiana, 1965.
43. G. Lakoff and M. Johnson, *Metaphors We Live By*, University of Chicago Press, Chicago, Illinois, 1980.
44. R. H. Thouless, The Tendency to Certainty in Religious Belief, *British Journal of Psychology, 26*, 1935.
45. K. Coles, A Theory of Everything, *The New York Times*, p. 28, October 18, 1987.
46. W. James, *Pragmatism*, Harvard University Press, Cambridge, Massachusetts, p. 97, (1907) 1975.
47. K. Popper, Normal Science and Its Dangers, in *Criticism and the Growth of Knowledge: Proceedings of the International Colloquim in the Philosophy of Science, London, 1965*, I. Lakatos and A Musgrave (eds.), Cambridge University Press, London, 1965.
48. P. Feyerabend, *Problems of Empiricism*, Cambridge University Press, London, 1981.
49. K. Kellen, Introduction to J. Ellul, in *Propaganda: The Formation of Men's Attitudes*, K. Kellen and J. Lerner (trans.), Random House, New York, 1965.
50. M. C. Jacob, *The Cultural Meaning of the Scientific Revolution*, Alfred A. Knopf, New York, 1988.
51. Galileo, *Two Chief World Systems*, S. Drake (trans.), University of California Press, Berkeley, California, (1632) 1967.
52. Galileo, *Discoveries and Opinions in Galileo*, S. Drake (trans.), Doubleday, New York, (1610) 1957.
53. H. Butterfield, *The Origins of Modern Science*, MacMillan, New York, 1957.
54. S. J. Gould, A Triumph of Historical Excavation, *The New York Review of Books*, p. 9, February 27, 1986.
55. F. M. Turner, Scientific Resolution, *Isis, 77*, pp. 508-511, 1986.
56. L. Laudan, *Science and Values*, University of Chicago Press, Chicago, Illinois, 1984.
57. M. Planck, *Scientific Autobiography and Other Papers*, F. Gaynor (trans.), New York, 1949.
58. T. Kuhn, *The Structure of Scientific Revolutions*, (2nd Edition), University of Chicago Press, Chicago, Illinois, 1970.
59. K. Wallace, The Substance of Rhetoric: Good Reasons, *Quarterly Journal of Speech, 44*, p. 242, 1963.
60. G. Stent, That Was the Molecular Biology That Was, *Science*, p. 363, April 26, 1968.

61. A. Novick, Phenotypic Mixing, in *Phage and the Origins of Molecular Biology*, J. Cairns, G. Stent, and J. Watson (eds.), Cold Spring Harbor Laboratory of Quantitative Biology, Cold Spring Harbor, 1966.
62. Plato, *Phaedrus*, P. Hackforth (trans.), Liberal Arts Press, New York, 1952.
63. I. Hacking, *Representing and Intervening*, Cambridge University Press, Cambridge, United Kingdom, 1983.
64. J. S. Mill, *On Liberty*, Hackett, Indianapolis, Indiana, (1859) 1978.

# DISCOURSE ON METHOD:
# THE RHETORICAL ANALYSIS OF SCIENTIFIC TEXTS

*Alan G. Gross*

## RHETORIC OF SCIENCE AS A DISCIPLINE

We readily concede that the law courts and the political forum are special cases of our everyday world, a world in which social reality is uncontroversially the product of persuasion. Many of us can also entertain a possibility Aristotle could never countenance: the possibility that the claims of science are solely the products of persuasion. We live in an intellectual climate in which the reality of quarks or gravitational lenses is arguably a matter of persuasion—such a climate is a natural environment for the revival of a rhetoric that has as its field of analysis the claims to knowledge that science makes.[1]

Rhetorically, the creation of knowledge is a task beginning with self-persuasion and ending with the persuasion of others. This attitude toward knowledge stems from the first Sophistic. In spirit, the *Rhetoric*, my master theoretical text, is also Sophistic, its goal "to find out in each case the existing means of persuasion." It is a spirit, however, that Aristotle holds firmly in check by limiting the scope of rhetoric to those forums where knowledge is unquestionably a matter of

---

[1] This paper is a pre/text in a real sense: it is, essentially, the first chapter of my book, *The Rhetoric of Science*. My work covers some of the same ground as Charles Bazerman's recent *Shaping Written Knowledge: The Genre and Activity of the Experimental Article in Science*, but differs markedly in aim. Bazerman wants to increase scientists' awareness of the rhetorical component of their communicative practice; I want to open up an intellectual space for a new subdiscipline, the rhetoric of science, a field that requires both a reinterpretation of the rhetorical tradition and the total abandonment of philosophical realism, two sides, I think, of the same coin. According to my program, science is rhetorically constituted to its core.

My work resembles Bazerman's in method: I too proceed by means of exigesis. In my book, I analyze an early Copernican treatise, Sprat's *History of the Royal Society*, Newton's *Opticks*, Darwin's *Notebooks*, contemporary papers in evolutionary taxonomy, biological chemistry, molecular genetics, and theoretical physics, peer review correspondence in both, the recombinant DNA controversy. To an extent, Bazerman and I cover the same textual ground; but, unlike Bazerman, I derive from my examinations of scientific texts the most radical of rhetorical conclusions. In spirit, my views are closest to those of that most radical of philosophers of science, Bruno Latour, whose *Science in Action* has for me the right iconoclastic ring. I also owe a substantial intellectual debt to the strong programme in sociology of science, especially, to Latour's one-time collaborator, Steve Woolgar.

But my debts are not entirely to the radical camp. I have strong ties to speech act theory, especially to the philosophy of society that Habermas derives from that theory: to Victor Turner's social anthropology; to analytical philosophy, especially Quine and Davidson; to ordinary language philosophy, especially Strawson and Grice; to the founder of sociology of science, Robert K. Merton, and to the founder of rhetorical studies, Aristotle.

persuasion: the political and the judicial. If scientific texts are to be analyzed rhetorically, however, this Aristotelian limitation must be removed the spirit of the first Sophistic must roam free.

Whether, after rhetorical analysis is completed, there will be left in scientific texts any constraints not the result of prior persuasion, any "natural" constraints, remains, for the moment, an open question. In the meantime, as rhetorical analysis proceeds unabated, science may be progressively revealed, not as the privileged route to certain knowledge, but as another intellectual enterprise, an activity that takes its place beside, but not above, philosophy, literary criticism, history, and rhetoric itself.

The rhetorical view of science does not deny "the brute facts of nature"; it merely affirms that these "facts," whatever they are, are not science itself, knowledge itself. Scientific knowledge is comprised of the current answers to three questions, answers that are the product of professional conversation: What range of "brute facts" is worth investigating? How is this range to be investigated? What do the results of these investigations mean? Whatever they are, the "brute facts" themselves mean nothing; only statements mean, and of the truth of statements we must be persuaded. These processes, by which problems are chosen and results interpreted, are essentially rhetorical: only through persuasion are importance and meaning established. As rhetoricians we study the world as meant by science.

## WHY A NEW DISCIPLINE?

Thirty years ago, the humanistic disciplines were more easily definable: historians of science shaped primary sources into chronological patterns of events; philosophers of science analyzed scientific theories as systems of propositions; sociologists of science scrutinized statements aimed at group influence [1, p. 43]. In the last two decades, however, the humanities have been subject to "a blurring of genres." As a result, "the lines grouping scholars together into intellectual communities . . . are these days running at some highly eccentric angles" [2, pp. 23-24].

David Kohn and Sandra Herbert on Darwin's Notebooks, and Gillian Beer on the Origin: are they writing intellectual history or literary criticism? Ian Hacking on gravitational lensing: is he doing philosophy or sociology? Arthur Fine on Einstein: is he doing philosophy or intellectual history? Are Steve Woolgar and Karin Knorr-Cetina studying the scientific paper from the point of view of sociology or rhetorical criticism? Is Evelyn Keller's work on Bacon epistemology, psychology, or literary criticism? Michael Lynch on laboratory shop talk: is he doing ethnomethodology or rhetoric of science?

These intellectual enterprises share a single methodological presupposition; all, to paraphrase Barthes, "star" their texts; all assume with Geertz that "the road to discovering . . . lies less through postulating forces and measuring them than through noting expressions and inspecting them" [2, p. 43]. To address Einstein's philosophy, Fine becomes a historian. Latour and Woolgar discover the intellectual structure of science, not through philosophical analysis, but through the ethnomethodology of the laboratory. Keller approaches Bacon's epistemology, not through the reconstruction of his arguments, but through the analysis of his metaphors; Beer treats the Origin less like an argument than like a novel by George Eliot or Thomas Hardy.

Rhetorical analysis describes what all of these scholars of science are doing; it defines the intellectual enterprise of workers as different in outlook and training as Gillian Beer and Steve Woolgar.[2] To such workers, the speculative knowledge of the sciences is a form of practical knowledge, a vehicle of practical reasoning, whose mark "is that the thing wanted is at a distance from the immediate action, and the immediate action is calculated as the way of getting or doing or securing the thing wanted" [3, p. 79]. The *Origin of Species* is speculative knowledge, certainly; from a rhetorical point of view, however, it is also practical knowledge, the vehicle by means of which Darwin attempted to persuade his fellow biologists to reconstitute their field, to alter their actions or their dispositions to act.

To call these intellectual activities rhetoric of science is only to register a claim already staked and mined; to view these apparently distinct enterprises as one is merely to make available to all a coherent tradition, a set of well-used intellectual tools.

Rhetoric of science differs from literature and science, a branch of study that also "stars" its texts. These branches differ in that the texts privileged by literature and science are traditionally literary; the science of an era is studied for its ability to illuminate the literature of that era: Katherine Hayles' *The Cosmic Web* trains the concepts scientific field theory on a set of contemporary novels, influenced by this theory. In contrast, rhetoric of science proposes by means of rhetorical analysis to increase our understanding of science, both in itself, and as a component of an intellectual climate. From this perspective, when Gillian Beer studies the impact of Darwin on Victorian intellectual life, she is doing, not literature and science, but rhetoric of science.

To say that a rhetoric of science views its texts as rhetorical objects, designed to persuade, is not to deny that there is an aesthetic dimension to science. From a rhetorical point of view, however, this dimension can never be an end in itself; it

---

[2] Of all workers, Bruno Latour makes the rhetorical orientation of his studies most explicit. In *Science in Action*, Latour places science among a web of activities that includes virtually every center of influence in human affairs. Persuasion, which constitutes each center, is also the binding force within the web.

is always a means of persuasion, a way of convincing scientists that some particular science is correct. In science, beauty is not enough: Descartes' physics is beautiful still, but it is not still physics.

## RHETORIC APPLIED TO SCIENTIFIC TEXTS

In a neo-Aristotelian rhetoric of science, the apparatus of classical rhetoric must be generally applicable; a formulation must be developed that is recognizably classical and, at the same time, a theory of the constitution of scientific texts. This is not to say that classical ideas of style, arrangement, and invention must be mapped point for point onto these texts. The notion is not that science is oratory; but that, like oratory, science is a rhetorical enterprise, centered on persuasion. Instead of searching for exact correspondences, we must, as we proceed, achieve a general sense that the categories of classical rhetoric can explain the observable features of scientific texts.

This task is made easier by the existence of a long tradition of rhetoric and rhetorical analysis. Classical rhetoric was never a unitary system, nor was the rhetorical tradition unified throughout its history: Aristotle, Cicero, Quintilian differ, as do Campbell, Whately, and Blair. Probably, if more texts survived, even more disagreement among classical authors would be evident.

But it is the continuities in the rhetorical tradition that are most striking, continuities that subsist generally throughout medieval and modern rhetoric. Writers still find arguments where classical orators found them; the organization of writing still owes a debt to classical ideas of arrangement; and rhetoricians still think of style in terms that are largely classical. When young people learn to write, they still learn what Quintilian taught.

The rhetorical analysis of science is made plausible by the close connection between science and rhetoric in the ancient world. Early Greek thought concerning the material world fluctuated wildly. To Thales, the fundamental substance was water; to Anaximenes it was air. To Heraclitus all was flux; to Parmenides, change was an illusion. To this *embarras de richesses*, there were two reactions. The first was to ensure the certainty of knowledge; this was the way of Plato and Aristotle. The second was to regard knowledge as human and changeable, as rhetoric; this was the way of the Sophists.

The problem that rhetoric of science addresses, then, was set early in the intellectual history of the West. And then as now, this problem cannot be addressed unless rhetorical analysis includes, not only the style and arrangement of science, but also those of its features usually regarded as unrhetorical, features commonly construed, not as rhetoric, but as the discovery of scientific facts and theories. From the rhetorical point of view, scientific discovery is properly described as invention.

Why redescribe discovery as invention? To discover is to find out what is already there. But discovery is not a description of what scientists do; it is a hidden metaphor that begs the question concerning the certainty of scientific knowledge. Discovered knowledge is certain because, like America, it was always there. To call scientific theories inventions, therefore, is to challenge the intellectual privilege and authority of science. Discovery is an honorific, not a descriptive, term; and it is used in a manner at odds with the history of science—a history, for the most part, of mistaken theories—and at odds with its current practice, a record, by and large, of error and misdirection. Invention, on the other hand, captures the historically contingent and radically uncertain character of all scientific claims, even the most successful. If scientific theories are discoveries, their unfailing obsolescence is difficult to explain; if these theories are rhetorical inventions, no explanation of their radical vulnerability is necessary.[3]

## Invention: *Stasis*

At any time, in any science, scientists must make up their minds about what needs to be explained, what constitutes an explanation, and how such an explanation constrains what counts as evidence. When scientists think about matters of explanation, they are deciding what it is to do science. In rhetorical terms, they are using stasis theory, an established part of invention, a set of questions by means of which we can orient ourselves in situations that call for a persuasive response. In courtroom arguments, we consider whether an act was committed (*an sit*); whether it was a crime (*quid sit*); whether the crime is justified in some way (*quale sit*). In the analysis of law, these stases have a central role; in the analysis of science, their centrality is equally apparent.

1. *An Sit.* In the sciences, what entities really exist? Does phlogiston? Do quarks? Before Einstein's papers on Brownian movement, the existence of atoms was in question; after, their existence was regarded as confirmed.
2. *Quid Sit.* Given that certain entities exist, what is their exact character? From antiquity, light has been steadily the subject of scientific scrutiny. Is light Aristotle's alteration in a medium, Descartes' pressure, Newton's particle, Young's wave (another alteration in a medium), or the zero-mass particle of quantum electrodynamics?

---

[3] In *The Social Basis of Scientific Discoveries*, Brannigan makes the analogous point that scientific discoveries are social constructions based on the novelty, validity, and plausability of candidate objects or events in the context of recognized programs of research.

3. *Quale Sit.* Even if the character of an entity or phenomenon remains roughly the same, the laws governing it may be radically different: the same law of refraction that is, for Newton, the result of deterministic forces acting on minute particles is, for Feynman, the product of probabilistic ones acting on zero-mass particles.

Particular scientific texts emphasize particular *stases.* Although Einstein incidentally established the physical existence of atoms, he was mainly concerned with the *quale sit* of Brownian motion. Papers in evolutionary taxonomy establishing a new species mainly support the *quid sit* of existence, but are also concerned with the *quale sit* of evolutionary theory. In every case, the *stases* focus the scientist's attention on a particular aspect of the problem before him: Newton and Descartes, for instance, were both concerned with the nature of white light, the *quid sit.*

There is a final *stasis* applicable to both rhetoric and science: whether a particular court has jurisdiction. Whether something is a scientific theory depends on who is doing the judging. Newton's formulation of the theory of light remained the same throughout his career. At first, it was rejected; its later acceptance depended, not on any alteration of the theory, but on a change of jurisdiction. In the first court of opinion, Newton was judged by rules of others' making; in the final court, the rules were Newton's own.

Jurisdiction is also important in adjudicating the relationships between science and society. At what point do decisions cease to be internal to a science? The Inquisition saw itself as an appropriate arbiter of all knowledge, including Galileo's scientific theories. In modern times, this determination is usually made by the scientific community. But even contemporary courts see themselves as the proper judges of the social impact of recombinant DNA research.

At any one time, in any one science, there are proper and improper ways to respond to the first three stases. For Aristotle, for example, the phenomena in need of explanation are those that naturally present themselves; what accounts for the motion of a stone released from the thrower's hand? This is a case of violent motion, a movement whose efficient cause is the application of a force to an object, overcoming its material cause, its *gravitas.* There is no violent motion without direct contact: a stone thrown in the air continues its motion after it leaves the thrower's hand only because of the impulsive power of the air directly behind it. The stone's initial trajectory is the formal cause of this violent motion. At the height of that trajectory, natural motion takes over; the stone begins to fall, seeking its natural place, the final cause of its motion. The material cause is again the stone's *gravitas,* its formal cause the downward trajectory itself, its efficient cause the distance from its natural place. For Aristotle, scientific explanation is essentially qualitative, according to the four causes; mathematics has no place in physics.

In his *Principia,* Newton escapes the Aristotelian an sit; in that work, he no longer takes as an explanandum the traditional topic of motion. For Newton, it is not motion, but change of motion that requires explanation. Motion itself—intuitively, the natural puzzle—is an *explanandum* no longer in the realm of science. Moreover, Newtonian explanations can be scientific that do not enumerate all four causes. The material cause of change of motion is largely bracketed, and its final cause assigned to theology. The efficient and formal causes are privileged, and the formal cause is given a mathematical interpretation: change of motion is explained according to strict mathematical relationships among such non-observables as force and mass. Such relationships permit quantitative solutions to problems in physics. Wherever possible, these problems have an experimental realization: in principle, though not in fact, Newton will assert only what he can observe under experimental constraints, or infer directly from controlled observation.

Because the presuppositions of Aristotle and Newton were opposed, because their notions of evidence and explanation seriously diverged, the sciences they created differed radically. Differently interpreted, the *stases* can lead, in fact, have led, to radically different conceptions of science. Since they precede science, the province of these interpretations cannot be science; their proper province is rhetoric.

## Invention: *Logos*

### The Common Topics

The common topics are a staple of classical rhetorical invention comparison, cause, definition—these and their fellows are the traditional places where rhetoricians can find arguments on any given topic. These same common topics are also an important source for arguments in science; in Newton, for example. In his *Opticks,* Newton defines a light ray twice. Early in this work, he provides a definition in terms of the observable: light behaves as if it were made up of tiny particles. Later, Newton defines light in terms of a hypothesis concerning the constitution of matter: light may actually consist of tiny particles. The difference in these definitions reflects a change in persuasive purpose. By means of the first definition, Newton hopes to persuade the skeptical scientist of the truth of his analysis of light; to agree, this scientist need not subscribe to Newton's speculative atomism. By means of the second, Newton hopes that this same scientist will seriously entertain atomism as a scientific hypothesis.

In Newton's optical works, the common topics are used heuristically as well as persuasively: Newton undermines Descartes' analysis of color by means of the topic of comparison: he contrasts Descartes' theory with his own incontrovertible experimental results. Concerned about the material constitution of light, he addresses the topic of cause: the sensation of light, he speculates, is evoked when its tiny particles impinge on the retina. In his presumption of the rectilinear propagation of light, he relies on the topic of authority: everybody since Aristotle has taken this as true.

In each case, we might say that Newton defines scientifically, compares scientifically. But in none of these instances is it possible to define a scientific sense for the common topics, qualitatively distinct from their rhetorical sense: these sources for arguments in science and rhetoric do not differ in kind.

## Special Topics

In addition to the common topics suitable to all argument, there are special topics that provide sources of argument for each of the three genres of speeches: forensic, deliberative, and epideictic. Forensic texts establish past fact; they are so named because their paradigm is the legal brief; their special topics are justice and injustice. Epideictic texts celebrate or calumniate events or persons of importance; their paradigms are the funeral oration and the philipic; their special topics are virtue and vice. Deliberative texts establish future policy; their paradigm is the political speech; their special topics are advantageous and the disadvantageous, the worthy and the unworthy.

Scientific texts participate in each of these genres. A report is forensic because it reconstructs past science in a way most likely to support its claims; it is deliberative because it intends to direct future research; it is epideictic because it is a celebration of appropriate methods. Analogously, scientific textbooks strive to incorporate all useful past science, to determine directions for future research, and to commend accepted methods. But science also has special topics of its own, unique sources for its arguments. Precise observation and prediction are the special topics of the experimental sciences; mathematicization is the special topic of the theoretical sciences. But there is considerable reciprocity. In the experimental sciences, mathematization is also a topic, and provides arguments of the highest status; and, in the theoretical sciences, at least by implication, arguments from mathematics are anchored in the special topics of prediction and observation. But are observation, prediction, and mathematization *topics?* Science is an activity largely devoted to the fit between theories and their brute facts; the better the fit, the better the science. Surely, observation, prediction, and mathematization are not topics, but means to that end. In prediction, the confrontation between theory and brute fact is at its most dramatic. Einstein's theory of general relativity forecast the never-before observed bending of light in a gravitational field; Crick's theory of the genetic code predicted that an otherwise plausible variant—the codon UUU—would never occur. Both predictions insisted on the participation of nature; nature, not human beings, would clinch the argument. Einstein's theory was confirmed by the bending of stellar light as measure during a total eclipse; Crick's was disconfirmed by the discovery of a UUU codon. In both cases, it seems, we have left rhetoric behind. We seem to be in direct contact with the brute facts as the criterion for theoretical truth: stellar photographs in the first case, instrument readings in the second.

But this line of argument fails: in neither case did the brute facts point unequivocally in a particular theoretical direction; in fact, in no scientific case do uninterpreted brute facts—stellar positions, testtube residues—confirm or disconfirm theories. The brute facts of science are stellar positions or testtube residues *under a certain description;* and it is these descriptions that constitute meaning in the sciences. That there are brute facts unequivocally supportive of a particular theory, that at some point decisive contact is made between a theory and the naked reality whose working it accurately depicts, is a rhetorical, not a scientific, conviction. Observation, prediction, measurement, and their mathematization: these are sources for the arguments in science in the same way—exactly the same way— that the virtuous is the source of arguments for the epideictic orator.

## Logic: The Structure of Argument

For Aristotle, scientific deduction differs in kind from its rhetorical counterpart. True, both are conducted according to the "laws" of thought. But rhetorical deduction is inferior for two reasons: it starts with uncertain premises, and it is enthymematic: it must rely on an audience to supply missing premises and conclusion. Since conclusions cannot be more certain than their premises, and since any argument is deficient in rigor that relies on audience participation for its completion, rhetorical deductions can yield, at best, only plausible conclusions. Rhetorical induction, reasoning from examples, is equally marked as inferior to its scientific counterpart because of its acknowledged inability to guarantee the certainty of its generalizations: examples illustrate, rather than prove.

Aristotle notwithstanding, rhetorical and scientific reasoning differ, not in kind, but only in degree. No inductions can be justified with rigor: all commit the fallacy of affirming the consequent; consequently, all experimental generalizations illustrate reasoning by example. Deductive certainty is equally a chimera; it would require the uniform application of laws of thought, true in all possible worlds, the availability of certain premises, and the complete enumeration of deductive chains. But of no rule of logic—not even the "law" of contradiction—can we say it applies in all possible worlds. Moreover, even were such universal rules available, they would operate, not on certain premises, but on stipulations and inductive generalizations. In addition, all deductive systems are enthymematic: the incompleteness of rhetorical deduction is different only in degree, not in kind, from the incompleteness of scientific deduction. No deductive logic is a closed system, all of whose premises can be stipulated; every deductive chain consists of a finite number of steps between each of which an infinite number may be intercalated [4, pp. 57-73]. Because the logics of science and rhetoric differ only in degree, both are appropriate objects for rhetorical analysis.

## Invention: *Ethos and Pathos*

### *Ethos*

Scientists are not persuaded by *logos* alone; science is no exception to the rule that the persuasive effect of authority, of ethos, weighs heavily. The anti-authoritarian stance, the Galilean myth canonizing deviance, ought not to blind us to the pervasiveness of ethos, the burden of authority, as a source of scientific conviction.[4] Indeed, the progress of science may be viewed as a dialectical contest between the authority sedimented in the training of scientists, an authority reinforced by social sanctions, and the innovative initiatives without which no scientist will be rewarded.

Innovation is the raison d'etre of the scientific paper; yet in no other place is the structure of scientific authority more clearly revealed. By invoking the authority of past results, the initial sections of scientific papers argue for the importance and relevance of the current investigation; by invoking the authority of past procedure, these sections establish the scientist's credibility as an investigator. All scientific papers, moreover, are embedded in a network of authority relationships: publication in a respected journal; behind that publication, a series of grants given to scientists connected with a well-respected research institution; within the text, a trail of citations highlighting the paper as the latest result of a vital and on-going research program. Without this authoritative scaffolding, the innovative core of these papers—their sections on Results, and their Discussions—would be devoid of significance.

At times, the effects of scientific authority can be stultifying: collective intellectual inertia blocks the reception of heliocentric astronomy for over a century; Newton's posthumous authority retards the reemergence of the wave theory of light. At other times, times perhaps more frequent, authority and innovation interact beneficially: consider heliocentric astronomy between Copernicus and Kepler, the theory of light between Descartes and Newton, the concept of evolution in Darwin's early thought; in each of these cases, we see the positive results of the dialectical contest between authority and innovation. These examples alert us to the fact that there is no necessary conflict between originality and deference. One of the persuasive messages of authority in science is the need to exceed authority; indeed, the most precious inheritance of science is the means by which its authority may be fruitfully exceeded: "Was du ererbt von deinen Vatern hasv Erwirb es, um es zu besitzen" [Goethe in Freud—5, p. 123].

---

[4] A broader definition of *ethos* is usual, one that includes matters of value which, for expository convenience, I place under *pathos* in the following section. Nothing significant rides on this idiosyncratic allocation.

At the root of authority within science is the relationship of master to disciple. To become a scientist is to work under men and women who are already scientists; to become a scientific authority is to submit for an extended period to existing authorities. These embody in their work and thought whatever of past thought and practice is deemed worthwhile; at the same time, they are exemplars of current thought and practice. In their lectures, they say what should be said; in their laboratories they do what should be done; in their papers, they write what should be written.

So long as science is taught as a craft, through extended apprenticeship, its routes to knowledge will be influenced by the relationships between masters and disciples. The modern history of heliocentricity is a progress from epicycles to ellipses. But this theoretical development is realized only through a chain of masters and disciples, surrogate fathers and adopted sons: Copernicus and Rheticus, Maestlin and Kepler. By this means, research traditions are founded, and the methodological and epistemological norms that determine the legitimacy of arguments are passed on as tacit knowledge.

An examination of the forms of authority within science reminds us that epistemological and methodological issues cannot be separated from the social context in which they arise: the early members of the Royal Society decided what science was, how it would be accomplished, how validated, how rewarded. But we need also to be reminded of another set of authority relationships: those between science and society at large. It was the paradoxical promise of early science that it would benefit society best when wholly insulated from larger social concerns. This ideological tenet becomes difficult to justify in an age of nuclear power and gene recombination. Justification is especially difficult when science converts its exceptional prestige into a political tool to protect its special interests, perhaps at the expense of the general interest. The recombinant DNA controversy is a case in point.

*Pathos*

Emotional appeals are clearly present in the social interactions of which science is the product. In fact, an examination of these interactions reveals the prominent use of such appeals: the emotions are plainly involved, for instance, in peer review procedures, and in priority disputes. Anger and indignation are harnessed in the interest of a particular claim; they are part of the machinery of persuasion. When science is under attack, in cases of proposed research in controversial areas, emotional appeals become central. The case of proposed research in gene recombination is a good example of the fundamental involvement of science in issues of public policy, and of the deep commitment of scientists to a particular social ideology.

In addition, the general freedom of scientific prose from emotional appeal must be understood, not as neutrality, but as a deliberate abstinence: the assertion of a value. The objectivity of scientific prose is a carefully crafted rhetorical invention, a non-rational appeal to the authority of reason: scientific reports are the product of verbal choices designed to capitalize on the attractiveness of an enterprise that embodies a convenient myth, a myth in which, apparently, reason has subjugated the passions. But the disciplined denial of emotion in science is only a tribute to our passionate investment in its methods and goals.

In any case, the denial of emotional appeal is imperfectly reflected in the scientific texts themselves. In scientific papers, the emotions, so prominent in peer review documents and in priority disputes, though far less prominent, are no less insistently present. In their first paper, Watson and Crick say of their DNA model that it "has novel features which are of considerable biological interest" [6, p. 737]. In his paper on the convertibility of mass and energy, Einstein says: "It is not impossible that with bodies whose energy-content is variable to a high degree (e.g., with radium salts) the theory may be successfully put to the test" [7, pp. 67-71]. In these sentences key words and phrases—"novel," "interest," "successfully," "put to the test"—retain their ordinary connotations. Moreover, in Watson and Crick, "considerable" is clearly an understatement: the topic is the discovery of the structure of the molecule that controls the genetic fate of all living organisms.

Our science is a uniquely European product barely three centuries old, a product whose rise depended on a re-focusing of our general interests and values. Its wellspring was the widening conviction that the eventualities of the natural order depended primarily, not on supernatural or human intervention, but on the operation of fixed laws whose preferred avenue of discovery and verification was quantifed sensuous experience. The ontological results aid this epistemological preference defined the essence of nature, and founded a central Western task: to control nature through an understanding of its laws. To this task, the specific values of science—such as the Mertonian norms of universalism and organized skepticism—are instrumentally subordinate. Equally subordinate are the values on which theory choice depends: simplicity, elegance, power. In such a view, the *ethos, pathos,* and *logos* are naturally present in scientific texts: as a fully human enterprise, science can constrain, but hardly eliminate, the full range of persuasive choices on the part of its participants.

### Arrangement

In science, the arrangement of arguments is given short shrift. It is hardly noticed, never taught; yet arrangement has always been important in modern science. Realizing its powerful effect, Newton cast his physics and recast his optics in Euclidian form. Indeed, during the three centuries of modern science, arrangement has become more, rather than less, important, more, rather than less, rigid; currently, form is so vital a component that no paper can be published that does not adhere closely to formal rules. In fact, the arrangement of scientific papers has become so inflexible that even experienced scientists occasionally chafe under its restrictive principles: results in this section, discussion in that. But when P.D. Medawar, a scientist of wide influence, put his Nobel weight behind a mild reform—putting the Discussion section first—his arguments were ignored, rather than answered [8, pp. 42-43].

Yet nothing is more artificial than the form of scientific papers. Experimental papers, for example, are less reports than enactments of the ideological norm of experimental science: the unproblematic progress from laboratory results to natural processes. It is of no consequence that such progress is far from unproblematic, or that the philosophical bases of this version of the scientific method have long been undermined. In experimental reports, arrangement behaves as a sacred given.

There is another aspect of arrangement, one even more central to the operation of science. Aristotle's decision to privilege the proofs of logic and mathematics, to except them from the province of rhetoric, was itself rhetorical, a decision in favor of certain arrangements, a choice that rested on their presumed correspondence to the laws of thought. It is a truism that logical and mathematical proofs are purely matters of syntax, of form, austere tributes to the power of pattern to evoke the impression of inevitability:

All A is B
All C is A
All C is Bqc

Like all syllogisms, *Barbara,* the paradigm syllogism of science, is sound only by virtue of its form, its arrangement. But so paradigmatic of absolute conviction have the forms of logic become, so binding has logical necessity seemed, that its force has been attributed to arguments in the natural sciences and, even, in the humanities: we speak of physical necessity and moral necessity, as if they and logical necessity were precisely analogous [8, pp. 193-260].

## Style

From the beginning, stylistic choices in modern science have been deliberately trivialized: in the words of Bishop Sprat, the first historian of modern science, its communications must "return back to the primitive purity and shortness, when men delivered so many things in an equal number of words." In such a program, the schemes and tropes of classical rhetoric are rigorously to be eschewed. Nouns stand for natural kinds; predicates for natural processes. Syntax, the structure of the sentence, is only the reflections of reality, the structure of nature.

Scientific style remains oxymoronic at its core: modest in its verbal resources, heroic in its aim—nothing less than the description of reality. Accordingly, trope like irony and hyperbole are barred; they draw attention away from the working of nature. Stylistic devices like metaphor and analogy cannot be condoned; they undercut a semantics of identity between words and things. Should scientific prose favor the active or the passive voice? This quarrel over schemes—over the appropriate surface subject of scientific sentences—masks essential agreement among the antagonists. Regardless of surface features, at its deepest semantic and syntactic levels, scientific prose requires an agent passive before the only real agent, nature itself. By means of its patterned and principled verbal choices, science begs the ontological question: through style its prose creates our sense that science is describing a reality independent of its linguistic formulations.

Despite these strictures, tropes like irony and hyperbole do appear regularly in scientific reports, belying the alleged reportorial nature of these texts, underscoring their true, persuasive purpose. Although the official view is that metaphor and analogy have only a heuristic function, that they wither to insignificance as theories progress, tropes are central to the scientific enterprise, and never disappear. In the *Origin of Species*, for example, a central argument is the analogy between artificial breeding and natural selection. This analogy was not abandoned as the theory matured; it was, instead, the means by which the theory has been maintained and extended. Analogy is also central to the whole enterprise of experimental science: laboratory experiments are scientifically credible only if there is a positive analogy between laboratory events and processes in nature.

In science, arrangement has an epistemological task, style an ontological task.

## ARISTOTELIAN RHETORIC UPDATED

To practice the rhetoric of science, then, is to make the Rhetoric the master guide to the exegesis of scientific texts. To perform this task effectively, the Rhetoric must be updated. The achievements of those squarely in the rhetorical tradition are the easiest candidates for incorporation into a neo-Aristotelian rhetoric of science. Of these, Perelman's work is most nearly central. His masterpiece, *The New Rhetoric*, written in collaboration with Madame Olbrechts-Tyteca, has as its strategic aim the rehabilitation of rhetoric as a discipline whose task is the analysis of persuasion in the humanities and the human sciences. Although Perelman does not deal with the natural sciences, the analysis of these is a plausible extension of the scope of his theory.

One central New Rhetorical concept useful in the analysis of science is the "universal audience," an ideal aggregate that can refuse a rhetor's conclusions only on pain of irrationality. Although the universal audience has been attacked as an ontological category, there is no disagreement that its assumption is a valid rhetorical technique [10, pp. 101-106]. Indeed, it is a technique essential to the sciences. The real audiences for papers in taxonomy and theoretical physics are vastly different in their professional presuppositions; nevertheless, all scientists attribute to imagined colleagues standards of judgment presumed to be universal: not in the sense that everyone judges by means of them, but in the sense that anyone, having undergone scientific training, must presuppose them as a matter of course.

There is a more sweeping, and more telling, criticism of *The New Rhetoric*, the accusation that Perelman and Olbrechts-Tyteca are seriously derelict in their philosophical duty: "One is never sure whether the authors are thinking of rhetoric primarily as a technique or primarily as a mode of truth. One wonders, too, what status the authors are claiming for the book itself" [9, p. 99]. This criticism of *The New Rhetoric* reminds us all to take an unequivocal stand on the epistemological status of our own inquiries. In my work, I view the techniques of rhetoric expounded by Perelman and Olbrects-Tyteca, techniques such as analogy, as the means by which we are persuaded that any mode of inquiry, including that of science, is a mode of truth.

A neo-Aristotelian theory of rhetoric should also be prepared to incorporate the results of relevant modern thinkers, those who purport to reveal through their work enduring qualitative patterns that undergird apparently unique verbal behavior. In rhetoric, Aristotle finds three persuasive appeals, three levels of rhetorical analysis. In an analogous fashion, the Russian formalist, Vladimir Propp, reveals that the dramatis personae of fairy tales exhibit thirty-one functions exercised in seven spheres of action; Sigmund Freud divides the mind's functions into ego, superego, and id; Jurgen Habermas analyzes speech acts by means of their relationship to their validity claims, to their communicative functions, and to reality.

The incorporation into a neo-Aristotelian rhetoric of science of views as divergent as those of Propp, Freud, and Habermas necessitates the abandonment of strong ontological claims. Aristotle's and Freud's psychologies cannot be incorporated into a single coherent theory. Additionally, an explanatory pattern in which we put great store may be, from another, equally legitimate, point of view, epiphenomenal, a symptom of the operation of purportedly more fundamental processes: Propp's patterns may be an effect of Freudian imperatives; Freudian imperatives, the result of the social dynamics of the upper middle class Viennese Jews who were Freud's contemporaries. Our choice among these patterns must be based, not on their relative truth, a judgment we cannot make, but on the amount each contributes to the understanding of the ways in which rhetorical processes constitute science.

## THE MOTIVE OF THE ENTERPRISE

In his *Crisis of European Sciences*, Edmund Husserl highlights the success of the natural sciences, a success to be contrasted with the general failure of reason in its task of improving the everyday world, the moral, mental, social, and physical space all human beings share. Husserl locates this failure in the rupture caused by the dualism of Descartes. Whatever its source, the breach between the world of science and our human world is real enough, and the task of reconciliation is as pressing today as it was for Husserl. Because it sees science wholly as a product of human interaction, rhetoric of science is a gesture in the direction of such reconciliation, an argument for the permanent bond that must exist between science and human needs.

The question of whether rhetorical analysis is appropriate and equal to so formidable a task arises, not as a consequence of any eternal truth or reasoned argument, but only as a result of the progressive narrowing and devaluation of rhetorical studies since Plato. It was Plato's successful attack on the Sophists that separated rhetoric from truth; it was the long authoritarian winter of the Roman Empire that limited rhetoric to its forensic and epideictic forms; it was the sterile intellectual reformulation of Ramus that reduced rhetoric to matters of style. That this narrowing was equally a degradation can be seen in such phrases as "mere rhetoric," "empty rhetoric."

Turning our backs on this past, we must engage in a systematic examination of the most socially privileged communications in our society: the texts that are the chief vehicles through which scientific knowledge is created and disseminated. We must argue that scientific knowledge is not special, but social, the result, not of revelation, but persuasion. In this way, we can see science as a permanent component of Husserl's life world, where it has its origin, and from which it must obtain all its purpose. In pursuit of this endeavor, rhetoric is the discipline of choice

## REFERENCES CITED

1. G. Markus, Why Is There No Hermeneutics of Natural Sciences? Some Preliminary Theses, *Science in Context,1*, pp. 5-51, 1987.
2. C. Geertz, *Local Knowledge*, Basic Books, New York, 1983.
3. G. E. M. Anscombe, *Intention* (2nd Edition), Cornell University Press, 1957.
4. P. J. Davis and R. Hersch, Mathematics and Rhetoric, *Descartes' Dream: The World According to Mathematics*, Harcourt, Brace, Jovanovich, 1986.
5. S. Freud, *An Outline of Psychoanalysis*, J. Strachey (trans.), W. W. Norton, New York, 1949.
6. J. D. Watson and F. H. C. Crick, A Structure for Deoxyribose Nucleic Acid, *Nature, 171*, pp. 737-738, 1953.
7. A. Einstein, *The Principle of Relativity: A Collection of Original Papers on the Special and General Theory of Relativity*, W. Perrett and G. B. Jeffrey (trans.), Dover, New York, (1923) 1952.
8. C. Perelman and L. Olbrechts-Tyteca, *The New Rhetoric: A Treatise on Argumentation*, J. Wilkinson and P. Weaver (trans.), Notre Dame University Press, Notre Dame, Indiana, (1958) 1971.
9. H. W. Johnstone, Jr., *Validity and Rhetoric in Philosophical Argument: An Outlook in Transition*, Pennsylvania State University, University Park, Pennsylvania, 1978.

# ADDITIONAL, UNNUMBERED REFERENCES

(in alphabetical order)

Bazerman, C., *Shaping Written Knowledge: The Genre and Activity of the Experimental Article in Science*, University of Wisconsin Press, Madison, Wisconsin, 1988.

Beer, G., *Darwin's Plots: Evolutionary Narrative in Darwin, George Eliot, and Nineteenth-Century Fiction*, Ark, London, 1985.

Beer, G., Darwin's Reading and the Fictions of Development, in *The Darwinian Heritage*, D. Kohn (ed.), Princeton University Press, Princeton, New Jersey, 1985.

Brannigan, A., *The Social Basis of Scientific Discoveries*, Cambridge University Press, Cambridge, United Kingdom, 1981.

Fine, A., *The Shaky Game: Einstein, Realism, and the Quantum Theory*, University of Chicago Press, Chicago, Illinois, 1986.

Gross, A., *The Rhetoric of Science: The Rhetorical Analysis of Scientific Texts*, Harvard University Press, Cambridge, Massachusetts, forthcoming.

Hacking, I., *Representing and Intervening: Introductory Topics in the Philosophy of Natural Science*, Cambridge University Press, Cambridge, United Kingdom, 1983.

Hacking, I., *The Making and Molding of Child Abuse: An Exercise in Describing a Kind of Human Behavior*, Harris Lecture, Northwestern University, May 7, 1986, 1-60 (unpublished).

Hayles, H. K., *The Cosmic Web: Scientific Field Models and Literary Strategies in the Twentieth Century*, Cornell University Press, Ithaca, New York, 1984.

Hayles, H. K., *The Politics of Chaos: Local Knowledge Versus Global Theory, Conference on Argument in Science: New Sociologies of Science/Rhetoric of Inquiry*, Iowa City, October 9-11, 1987.

Herbert, S., The Place of Man in the Development of Darwin's Theory of Transmutation, Parts 1 and 2, *Journal of the History of Biology, 7 and 10*, pp. 217-258, 1974 and pp. 155-227, 1977. E. Husserl, Edmund. *The Crisis of European Sciences and Transcendental Phenomenology*, D. Carr (trans.), Northwestern University Press, Evanston, Illinois, 1970.

Keller, E. F., *Reflections on Gender and Science*, Yale University Press, New Haven, Connecticut, 1985.

Knorr-Cetina, K. D., *The Manufacture of Knowledge: An Essay on the Constructivist and Contextual Nature of Science*, Pergamon, Oxford, United Kingdom, 1981.

Kohn, D., Theories to Work By: Rejected Theories, Reproduction, and Darwin's Path to Natural Selection, *Studies in History of Biology, 4*, pp. 67-170, 1980.

Latour, B., *Science In Action*, Harvard University Press, Cambridge, Massachusetts, 1987.

Lynch, M., *Art and Artifact in Laboratory Science: A Study of Shop Work and Shop Talk in a Research Laboratory*, Routledge & Kegan Paul, London, 1985.

Medawar, P., *Pluto's Republic*, Oxford University Press, 1984.

Woolgar, S., Discovery: Logic and Sequence in a Scientific Text, in *The Social Process of Scientific Investigation*, K. D. Knorr, R. Krohn, and R. Whitley (eds.), D. Reidel, Dordrecht, 1981.

# CHAPTER 3

# Social Constructionism

Social constructionism is the understanding that our world of knowledge and experience is socially negotiated and ratified and therefore constructed rather than absolute, pre-existent, or simply given. This constructedness pertains both to everyday reality and to the specialized world of science and technology. Thus scientific and technical knowledge is seen more as invented and shaped than discovered or represented because nothing is ever known absolutely directly. Social constructionism is closely related to the other topics in this volume. It is concerned with the fact of the social constructedness of knowledge, meaning, and the world, while rhetoric deals with the manner of this constructedness. The feminist advocacy of a gender-fair social order is an instance of social constructionism, especially in emphasizing the cultural determinants of behavior (e.g., Lorber and Farrell's *The Social Construction of Gender* [1]). Ethics is obviously socially constructed in that it involves the human criticism of human behavior.[1]

Though social constructionism is a development of contemporary philosophy and sociology, it carries profound implications both for composition studies generally and for technical and scientific communication specifically.[2] Chief of

---

[1] H. W. Simons offers a somewhat different relationship among these topics, including as "rhetorics of inquiry" social constructionism, various ethnomethodologies, and feminism [2, p. 7]. Whether social constructionism subsumes rhetoric, is subsumed by it, or is collateral with it is not crucial for our purposes. I will consider them simply as allied.

[2] Strictly speaking, the idea of social constructionism is not new. In classical Greece, an on-going debate opposed *nomos* as human laws which were mutable, contingent, and based on convention against *physis* as the immutable, absolute, pre-existent lawfulness of nature and essences. Kerferd is a thorough reference on this debate, while Nussbaum is a passing though incisive reference [3, 4]. See von Glaserfeld for a concise philosophical history of social constructivism from the Pre-Socratics through Vico to recent times [5].

these implications is the reconceptualization of the role of the technical and scientific communicator. Rather than being seen as only a passive relayer of received, privileged, and already-established knowledge, the technical communicator is now seen as potentially the developer, shaper, even generator of new knowledge.

In this chapter I will present an overview of social constructionism—its definition, its origins, and the views of two prominent philosophers on it. Next I will review the literature relating social constructionism to technical and scientific communication. Then I will present three key theoretical issues it faces. Finally I will introduce the two reprinted articles.

## OVERVIEW OF SOCIAL CONSTRUCTIONISM

### Definition and History

Social constructionism for our purposes means the pushing into the foreground the social factors by which knowledge, meaning, and our world are defined, whether consciously or not. These factors include the social conceptualization, negotiation, and ratification of such matters, especially within a scientific discipline. It considers, too, concepts, attitudes, beliefs, and the values that determine what is taken as fact, in T. Kuhn's research on paradigm shifts in science for example [7]. It also necessarily involves considerations of power, politics, and culture criticism.

In practice, social constructionism involves both criticism and affirmation. On the one hand, it affirms the validity of new sorts of knowledge and the empowerment of the historically disempowered, while on the other hand it challenges the exclusive privilege historically accorded scientific and technical knowledge. The knowledge and criticism of non-specialists, such as minorities and women, from this perspective has equal claim to potential social ratification as does the knowledge of specialists. Such groups have long been excluded from, say, debates about governmental expenditures for astronomy, when what was simultaneously though tacitly at stake was expenditures for child health care or shelter for the homeless.[3]

For a succinct summary of the principles of social constructionism, C. R. Miller's landmark essay on the humanistic rationale for technical writing serves well (though Miller does not use the term "social constructionism") [8]. Miller opposes the historical but erroneous transparency view of language as being suspicious of both language and humanity and fostering an invidious impersonalness. She argues instead that "reality" is not absolute but is conditional on a

---

[3] My representation of social constructionism is rather abbreviated. For a more thorough treatment regarding science, I recommend A. Pickering's *Science as Practice and Culture*. Pickering adopts an instrumentalist conception of science by which scientific theories become "closed systems" which confirm themselves and which are incommensurable with the instrumentalities of other theoretical systems [6].

knower and a community. Miller argues for a "communitarian" view which recognizes the fundamental interpretive indeterminacy and social contingency of a great deal (if not all) of technical knowledge.

## Social Constructionism, Positivism, and Science

Social constructionism is often defined by contrast to positivism in science (though the representativeness of positivism for all of science can be problematic). Positivism as a movement in philosophy and science sought the absolutely true, what could known positively and what was impervious to, even above, human opinion. By seeking what was above the mutability and fallibility of human opinion, however, it is thought by many critics to have demonstrated a profound disdain for our very humanity. For brief histories of this suspicion back to the seventeenth century, see Miller, Halloran, and Halloran and Whitburn (though this thinking is actually much older) [8-10]. Unfortunately, one of the legacies of positivism has been a deep suspicion of language for its chronic fuzziness, its shifting shape, and its inherent social contingency.

The notion that our world of experience and habitation is socially shaped is not necessarily radical, being a commonplace in literature studies for example. What is novel about contemporary social constructionism, however, is its origins, its authority, and its stance in direct opposition to the traditional understanding of science. It originated within the scientific community itself through medical scientists such as Fleck, sociologists such as Merton, anthropologists such as Mead and Geertz, philosophers of science such as Kuhn and Feyerabend, and sociologists of science such as Berger and Luckmann, Knorr-Cetina, and Latour and Woolgar [11-14, 7, 15-18].[4] The surprise is thus not social constructionism per se but that what is often taken as a radical confrontation with traditional science has come paradoxically from within science and carries its authority.

### Developments from Philosophy

In philosophy, R. Rorty in particular has challenged basic assumptions about foundational authority [20]. Every aspect of the search for authority in objective truths is doomed, he holds, to circularity and failure.

All we have, Rorty says, is social "solidarity." This solidarity sanctions a particular system or web of beliefs about what will be taken as reality [21].[5] This system is among a universe of possible systems, any of which theoretically could

---

[4] See Gergen for a concise, perceptive history of social constructionism as a movement in modern psychology and its connections to modern philosophy [19].

[5] See Quine and Ullian's *The Web of Belief* for an accessible discussion of reality as an interwoven web of beliefs rather than as a collection of positivistic facts [22].

have been selected for ratification. Alteration to this system occurs when, Rorty says, we collectively "bump our heads" against something dissonant with our system. This dissonance then impels the adjustment of the system of beliefs so as to accommodate this new datum. This externality does not of itself define our reality but rather our reality is defined by the entire interconnected system of beliefs, many of which have little direct connection to externalities (our beliefs about gods, say, or about scientific causality). To Rorty's mind, furthermore, viewing knowledge as always socially constructed naturally compels the discussion of ethics in science and technology.

P. Feyerabend, a philosopher of science, resonates with Rorty in holding science to be only one among a universe of other, equally valid alternative world views (e.g., ethics, religion, or Zuni culture) [15]. In addition, Feyerabend traces both the psychological as well as the social contingency of knowledge. He shows, for instance, that Galileo's claim that he observed moons revolving around Jupiter was an observation conditioned by his prior conceptualization that the earth was not the center of the universe. That is, theory actually precedes and *yields* observations, not the other way around, as is commonly thought.

## LITERATURE REVIEW

### General

The general influence of social constructionism on both the academic and the non-academic worlds has been significant. In the academic world, there has already been a major impact through the work of K. Bruffee and others in composition studies, for example [24]. Indeed, a whole new academic field, Science, Technology, and Society, has emerged specifically to explore the social aspects of science. S. E. Cozzens and T. F. Gieryn, for instance, review the evolution of theories of the sociology of scientific knowledge (SSK) [25]. R. Hagendijk, similarly, reviews the history of constructivism in the sociology of knowledge, from a focus on the individual researcher to the normative social principles acting on the research [26].

In the non-academic world, the ripple of these developments is also felt. C. Waddell, for instance, has documented the assertion of authority by the citizens of Cambridge, Massachusetts to control genetic engineering research in their environs, encompassing MIT and several other research institutions [27]. Prior to this, the authority to decide the permissibility of research had rested, for most practical purposes, within scientific community itself on the basis that it alone could fully grasp the significance of, need for, and potential dangers of their research. Now, however, non-specialists are asserting their authority to criticize, limit, and control both the development and the application of specialized knowledge.

## Composition and Communication Generally

The broadest works on social constructionism in composition and communication are those of LeFevre and Bruffee. LeFevre's *Invention is a Social Act* is a scholarly, highly regarded study of the history of the development of knowledge through social means [29]. It reveals that the scientist never discovers in isolation and that the writer never rhetorically "invents" his or her writing in isolation: knowledge is always social. The book is accessible and covers both scientific and technical fields and literary and composition fields in one work.

Bruffee's *College English* article also presents a broad, accessible survey of the history of social constructionism [24]. I should caution, however, that the wholesale translation from composition to technical and scientific communication is not clearly warranted. Composition often entails a personalistic authority and often does not have to face empirical verification. In light of such constraints, composition should perhaps be considered only as a preliminary to technical and scientific communication, not as a fully collateral activity.

## Technical and Scientific Communication Specifically

In the last few years, several books relevant to social constructionism in scientific and technical communication have been published, most notably C. Bazerman's *Shaping Written Knowledge* [28]. As convenient introductions to Bazerman and others, two review articles are available. I recommend Journet for its conciseness and simplicity [30]. I recommend Miller (reprinted later), on the other hand, for its sophistication and rigor and for its criticism [31]. Miller additionally discusses key issues facing the further application of social constructionism.

L. Odell's *Writing in Nonacademic Settings* includes four essays relating directly to technical writing [32]. L. Faigley's Nonacademic Writing: The Social Perspective emphasizes the social and cultural context in which communication occurs, the "discourse community" [33].[6] It deals also with the social conflicts, tensions, and interests at work in collaboration among group writers (see elsewhere Lay and Karis on collaboration [35]). L. Odell's *Beyond the Text: Relations Between Writing and Social Context* points out that the organizational context amounts to a culture with its own particular attitudes, values, goals, and systems of knowledge in which members participate [36]. Odell argues that teaching professional and technical writing must help in the general process of inquiry itself, a social process that begins even before a draft is conceived. J.

---

[6] See, however, T. Kent for a critique of "the very idea of 'discourse community,'" in particular the role of communication in spanning disparate communities [34].

Paradis, D. Dobrin, and R. Miller's *Writing at Exxon ITD: Notes on the Writing Environment of an R&D Organization* note that the writing of technical communication practitioners is much more socially involved than they themselves realize [37]. Such communications serve a variety of important social purposes, only some of which are consciously acknowledged by practitioners. C. R. Miller and J. Selzer's *Special Topics of Argument in Engineering Reports* reveals that many of the key writing "topics" [from the *topoi* of Greek rhetoric] of specialized fields are themselves specialized (in addition to general rhetorical topics with which many writers are already familiar) [38]. Thus much of the key work of technical and professional communication cannot be understood, mastered, or taught without an intimate familiarity with that specialized field.

Another book, *Professional Communication: The Social Perspective*, edited by N. R. Blyler and C. Thralls, contains two essays particularly germane to our topic [39]. Thralls and Blyler's *The Social Perspective and Professional Communication: Diversity and Directions in Research* reveals that "the social perspective" has developed into three distinct approaches: the social constructionist (defined narrowly), the ideologic, and the paralogic hermeneutic [40]. Social constructionism takes the identity, purposes, processes, and products of the writer to be fully determined socially. The ideological approach deliberately focuses on political issues and explicitly advances a political agenda. The paralogic hermeneutic approach rejects "the very idea" that language, knowledge, and rhetorical contexts are only socially defined, sharply delineated, or strictly determined.

T. Kent's essay in the same book, *Formalism, Social Construction, and the Problem of Interpretive Authority*, offers a more complete exposition of his paralogic hermeneutic view [41]. Kent holds that communication is much more interactive, chancy, and contingent than conventional social constructionism would have it. Meaning is so contingent on unpredictable social and personal variables that it cannot be completely anticipated and so is fundamentally indeterminate. The upshot is to expand the denotation of "social constructionism" to attenuate the boundaries of discourse communities while disprivileging their authority.

Among journal articles dealing specifically with technical and scientific communication, D. Winsor offers two treatments of different depths. *Engineering Writing/Writing Engineering* introduces social constructionism and discusses its potential application in general terms, emphasizing the conceptual reorganizations involved [42]. *The Construction of Knowledge in Organizations: Asking the Right Questions about the Challenger,* on the other hand, assumes a familiarity with social constructionism and examines the problematic application of it to the investigations of the Challenger disaster, at once a more particular and a more sophisticated treatment [43].

My own article about the Challenger disaster (reprinted later) shows how specific events leading up to the disaster demonstrate the social construction of knowledge in powerful, real terms [44]. I refer, however, to the "social contingency of meaning" rather than to the "social construction of knowledge" to avoid the misleading connotations of "construction." The article concludes by indicating that the raw, objective data themselves are rarely the true subject of technical communication; instead the subject is primarily the meaning, significance, and interpretation attached to these data—the subjective and social implications of the objective.

C. Bazerman's review essay, *Scientific Writing as a Social Act: A Review of the Literature of the Sociology of Science*, offers a broad overview of the history, philosophy, politics, and sociology of science relating to technical and scientific writing through about 1982 [45]. Bazerman is particularly helpful in tracing the intellectual history of the idea of the social construction of scientific and technical knowledge, and the vital interpretative importance of understanding the entire personal and social context in which science is conducted. To fully understand writing, Bazerman explains, the entire interrelated social web of beliefs, interests, and activities must be understood.

## THREE KEY THEORETICAL ISSUES

Though social constructionism is already established as an important intellectual development of our times, it has yet to evolve fully. Three key issues, discussed below, face the continued growth and application of social constructionism. The first has to do with how strongly to take it as a movement and as an intellectual stance. The second has to do with the authority of it, both reflexively as itself a social construct and as an implicit critique of traditional views of science. The third has to do with the indefiniteness of the society referred to in "social constructionism," which must be clarified in any use of the term.

### Radical versus Conditioned

The first issue is whether science is totally rhetorical and socially constructed or not. This issue is related to the traditional objection to philosophical idealism, namely that the "thing-in-itself" can never be known directly. (This is the same issue Miller raises in her review, reprinted later, of Bazerman's work on the rhetoric of scientific writing.)

This issue can be represented as two fundamental questions: Are our social constructions totally unconstrained, free to develop in whatever direction or form we choose to ratify? [7] Or, on the other hand, do we have a great deal of latitude in our constructions though we are constrained at some ultimate points by

---

[7] As we saw in Chapter One, *Rhetoric of Science*, A. Gross takes the radical stance that there are no constraints to the rhetoricity of science, or, put differently, science is rhetorical "all the way down."

an external, non-human environment? One implication is to wonder about the ethos of the technical communication instructor if he or she rejects radically, *in toto*, the authority of empirical science. Alienating our audience of scientists and technicians, after all, would ultimately be counterproductive.

These questions remind me of the elegant explanation Einstein offered about his relativity theory [46]. In trying to describe to a lay audience his understanding of the extent of our universe, Einstein juxtaposed and contrasted "limit" and "bound." He explained that the universe is finite and limited yet it is unbounded. That is, our observations are limited by finite parameters but are nevertheless without bounds. Einstein used the analogy of two-dimensional beings living on the surface of a sphere. Though the sphere is of finite and constant radius, the two-dimensional beings can roam the surface of the sphere in any direction indefinitely, without end. This is how I conceive of the empirical constraints on social constructionism—allowing indefinite latitude of construction within nevertheless definite constraints.[8]

Part of the reason for these question is that social constructionism is often represented dualistically as being radically opposed to empirical, objectivist science. I think, however, that such a dualistic approach is unnecessary. To illustrate how both empirical, objectivist science and social constructionism might peacefully, usefully co-exist without insisting on the invalidity of the other, consider the history of physics.

Less than one hundred years ago, there was a spirited clash between two major views of physics. Physics with sharp edges encompassed Newtonian mechanics, strict determinism of an all-or-none sort, and the particulate view of matter. Physics without sharp edges encompassed quantum mechanics with only probabilistic determinacy regarding the motion of matter as waves rather than as particles. Einstein, however, vehemently rejected Planck's probabilistic quantum mechanics without sharp edges, arguing that "God does not play dice."[9]

The initial clash evolved, however, to the present state of co-existence in which what amounts to different cultures or world-views within physics are recognized as equally and simultaneously valid. Newtonian mechanics and particle physics is seen now as only a special case of the more general quantum, wave physics.

---

[8] E. von Glaserfeld makes observations in a similar vein in describing the relation between "reality" and our cognitive constructions [5]. He explains that "reality" does not *cause* our cognitive constructions but only operates to select negatively those which are not viable. Thus von Glaserfeld posits the relation of knowledge to "reality" as being a severe constraint which nevertheless tolerates a wide range of existences.

[9] T. Kuhn, though for some a principal exponent of social constructionism, interestingly maintains constraints on his view of constructionism. In *The Natural and the Human Sciences,* for example, he maintains a fundamental distinction between the natural and the human sciences, holding that a certain stability or permanence of reference (which does not necessarily imply essentialism) is the basis for the distinction [47].

When one is interested in predicting eclipses, one turns to Newtonian physics for an easy answer. When, on the other hand, one is interested in sub-atomic interactions, one turns to quantum mechanics. Thus, the investigator shifts world-views to suit the investigation rather than being inflexibly bound always to one only.

An additional illustration of the flexible choosing of world view among co-existing alternatives that avoids simplistic dualism is E. Fox Keller's views on the controversies between sex and gender, and between biology and constructionism (see also discussion of Keller in the chapter on feminism) [48]. Keller acknowledges the empirical, objective reality of biological sex as genotype and as phenotype, but at the same time recognizes the limitations of that entire sphere of knowledge.

For questions about what women and men can and should do in our society, Keller turns to the sphere of social constructionism and gender. For the genetic determinants of different morphologies, she turns instead to the sphere of science and biology. One of the key values of Fox Keller's perspective is precisely that she recognizes the interrelated and simultaneous co-existence of the two spheres and the area where they interface. Biology, she holds, puts limits on the range of possible constructions, while social construction defines the broad range of social expression of genotype and phenotype, the sociotype, if you will.

## Nature of Social Constructionism Itself

The second issue is the nature of social constructionism itself. Social constructionism as a field of study is an artifact of science, itself a sort of science. To the extent that it critiques the claims of absoluteness of traditional scientific knowledge, one must wonder about its own authority. Would it equally be a mistake to take the proclamations of social constructionism as absolute and unequivocal as it was to take those of traditional science?

Taking this idea further, we should recognize that social constructionism is itself socially constructed and avoid the fallacy of self-exclusion with respect to social constructionism. We should, that is, remain mindful of the social and historical contexts of social constructionism and the particular interests it serves as an ideology, an -ism. As Berger and Luckmann observed in formulating their theory of social constructionism, a characteristic of the social construction of reality is that it is so taken for granted as obviously correct that no question about its authority ever comes to mind.

## The Operative "Society"

The third issue is the "society" in question when one speaks of "social construction." This issue arises from the loose usage of "social constructionism" in some of the literature and can be clarified, by way of analogy, by reading T. Kent on "the very idea of a discourse community" [34]. The society, its definition, constitution, and membership is rarely discussed, as though we all know or should know what is being talked about.

But looking at a particular instance such as the Challenger disaster reveals that there is no obvious referent of "society" or "social." Would "social" refer to NASA; or to Morton Thiokol, Inc. (MTI); or NASA plus MTI? Would it include or exclude the astronauts (a point emphasized by both the Presidential and Congressional investigations) [49, 50]? Would it refer to the managers at either institution, or to the engineers, or to the two groups together? Or would it refer to the Federal government as a whole or to the American public as well (as the Presidential investigation held)? This question is important because it tells us who is doing the constructing and what are the bases for ratifying the construction.

Even more importantly (continuing the analogy from Kent), there is the question of the social identification of any one individual. Does a person belong to or identify with only one society at time, exclusively, or does his or her identification shift continually and amorphously? Can multiple identifications be held simultaneously? Who indeed does the identifying, the person or an outside critic?

On this issue, the Challenger disaster is again illustrative. A crucial item of testimony before the Presidential commission investigating the disaster came from Robert Lund, Vice-President of Engineering at MTI. He testified that during the crucial debate between MTI engineers and managers the night before the launch, during which engineers refused to certify the safety of the flightworthiness of the booster, he was asked by Jerald Mason, Senior Vice-President of Operations, ". . .to take off his engineering hat and put on his management hat" [49, vol. V, p. 94]. With this alteration of "society" of identification came a reversal of the decision against launching. Not only did Lund's identification shift radically from engineer to manager, but the person instigating the alteration and making the identification was not Lund himself. Thus his identification and entire frame of reference was socially constructed.

Unless clarifications of issues such as these are made with regard to any particular instance, saying knowledge is "socially constructed" of itself is not informative. I do not mean by raising these issues to minimize the already-significant contributions of social constructionism to the understanding of human communication but only to suggest the rigor required for the careful, fruitful application of it to any particular instance. Unless such questions are answered, our discussions will be prone to emptiness or circular self-confirmation, jeopardizing the realization of the great promise of this philosophical perspective.

The difficulties which I have mentioned above are echoed by other critics, who suggest other difficulties as well. The theoretical crux Miller identifies in the article reprinted below is perhaps the principal difficulty in applying social constructionism to technical communication; it amounts to the question I raised earlier of radical versus conditioned constructionism. The broad theoretical significance of this issue even within philosophy of science itself is indicated by

Hiley et al.'s *The Interpretive Turn*, a collection of essays among philosophers of science on just this point [51].

Of more concrete and immediate relevance is the example of environmental movement. N. Evernden explains in *The Social Creation of Nature* that we assume that nature is a given, absolute and separate from our beliefs [52]. He shows, however, that "nature" is always socially defined and contingent on social, political, economic factors, though it is used in ways which obscure this contingency. "Nature," that is, self-reflexively indicates a conceptualization which supports our prior expectations and values. As a result, Evernden shows, arguments about "nature" can be used to support apparently opposite positions such as radical conservation or managed use of resources.

The debate about the range of applicability of social constructionism is yet to be resolved. The ramifications of social constructionism as itself a construct and, at times, an ideology, have yet to be worked out. It is nevertheless clear that at least a limited, conditioned social constructionism is generally valid, useful, and consonant with a humanistic understanding of science and technology as social enterprises. Conditioned constructionism, additionally, offers an important venue for making the practice of science and technology more accountable and responsive to society in generally.

## REPRINTED ARTICLES

### Dombrowski

In *Challenger and the Social Contingency of Meaning: Two Lessons for the Technical Communication Classroom,* I illustrate the applications of social constructionism to actual events, showing that knowledge and meaning, even of a highly technical nature, do not exist absolutely and separately from our conceptualizations and constructions [44]. I selected the Challenger disaster because of the clear social contingency and construction of meaning revealed in the voluminous testimony and evidence, and because viewing the disaster itself was a vivid, personal experience for nearly all Americans. Thus its lessons are not abstractions but meaningful "reality".

The raw material of meaning, the hard data on the charring of the O-ring seals, was unquestioned. What was questioned and heatedly argued, however, was the meaning of these data, and it is the history of the evolution of this socially contingent meaning that I trace using the testimony before both the Presidential commission and the Congressional committee.

The lessons I draw directly oppose the view of technical communication emphasizing objectivity, factuality, and information. From the perspective of the impact of the social forces acting in the Challenger episode, the objective data themselves can appear almost trivial. Everyone agreed on what the raw data were, yet there was violent disagreement as to what they represented; that is, the data did not determine strictly the representation of them. The lessons are simple but of profound significance. "The thing, the charred seal, is itself of little interest or even meaning in itself for technical communication. Only when we begin to grapple with such questions as—what does this mean? how is it to be interpreted? and what are we to do in light of this?—does a valid rhetorical object emerge for technical communication" [44, p. 82]. Thus technical communication is not principally the impersonal relaying or reporting of information but as much the distinctly personal and social activity of shaping meaning and molding interpretations.

## C. R. Miller

Miller offers an excellent review of the ideas of social constructionism and the rhetoricity of science [31]. Though Miller reviews three books, her discussion of Bazerman's is the most germane to this chapter.

Miller begins by exploring the root question of the connection between rhetoric and science. Historically, she says, arguments of two principal sorts have been offered against a connection: from essences and from circumstances. Both arguments have been effectively refuted by postmodern philosophy, she explains.

Miller lauds Bazerman's *Shaping Written Knowledge* as "perhaps the first sustained, coherent, historical study of [the rhetoric and social construction of] scientific discourse" [31, p. 110]. She identifies in Bazerman's account of rhetoric in scientific writing, however, a crux which also applies to social constructionism in general (see also my earlier distinction between radical and conditioned social constructionism). Bazerman, Miller says, because he sees "science as an enterprise committed to 'empiricism,'" grapples with the problem of reference entailed in empiricism. Miller explains that Bazerman abjures the radical social constructionist position represented by "postmodern sign theory" [31, p. 113]. In its place Bazerman develops "a social-constructivist approach to reference, which turns reference into accountability" [31, p. 113]. Thus Bazerman attempts to transmute reference into rhetoric.

Miller acutely recognizes the difficulty with this reconceptualization, suggesting that it "simply begs the question of reference" [31, p. 113]. That is, Bazerman does not so much transmute reference into rhetoric but renames "reference" as "accountability" while maintaining its essential substance: correspondence to material reality. Bazerman's reconceptualization amounts to what I have called "conditioned social constructionism," which agrees with Bazerman in holding that social constructionism is ultimately constrained by "empirical experience" [31, p. 113]. Miller, then, sees Bazerman not so much as revealing the essentially rhetorical nature of science but rather as redefining it while retaining the idea of reference. Miller concludes that though Bazerman retains the root distinction and merit of science as its grounding in empirical experience, he nevertheless "has outlined a new approach to the problem of reference, . . . one with a rhetorical flavor, given by its reliance on community, convention, and persuasion" [31, p. 114].[10]

Bazerman therefore seems to identify the limits of social constructionism in an external, empirically studied reality. And though Miller does not say so explicitly, she appears to concur, in characterizing Bazerman's rhetorical project as "sophistic," that is, pragmatic. Such sophism is rooted in the origins of humanism such as Protagoras's dictum: Man is the measure of all things.

## REFERENCES

1. J. Lorber and S. A. Farrell, *The Social Construction of Gender*, Sage Publications, Newbury Park, California, 1991.
2. H. W. Simons (ed.), *The Rhetorical Turn: Invention and Persuasion in the Conduct of Inquiry*, University of Chicago Press, Chicago and London, 1990.
3. G. B. Kerferd, *The Sophistic Movement*, Cambridge University Press, Cambridge, England, 1981.
4. M. Nussbaum, Aristophanes and Socrates on Learning Practical Wisdom, *Yale Classical Studies, XXVI*, pp. 43-98, 1980.
5. E. von Glaserfeld, An Introduction to Radical Constructivism, in *The Invented Reality*, P. Watzlawick (ed.), W. W. Norton, New York, 1984.
6. A. Pickering, *Science as Practice and Culture*, University of Chicago Press, Chicago and London, 1992.
7. T. Kuhn, *The Structure of Scientific Revolutions*, University of Chicago Press, Chicago, 1970.
8. C. R. Miller, A Humanistic Rationale for Technical Writing, *College English, 40*:6, pp. 610-617, 1979.
9. S. M. Halloran, Eloquence in a Technological Society, *Central States Speech Journal, 29*, pp. 221-227, 1978.
10. S. M. Halloran and M. D. Whitburn, Ciceronian Rhetoric and the Rise of Science: The Plain Style Reconsidered, in *The Rhetorical Tradition and Modern Writing, Modern Language Association*, J. J. Murphy (ed.), New York, 1982.
11. L. Fleck, *Genesis and Development of a Scientific Fact*, T. J. Trenn and R. K. Merton (eds.), F. Bradley and T. J. Trenn (trans.), T. S. Kuhn (foreword), University of Chicago Press, Chicago, 1979.
12. R. K. Merton, *The Sociology of Science*, University of Chicago Press, Chicago, 1973.
13. G. H. Mead, *Mind, Self, and Society from the Standpoint of a Social Behaviorist*, University of Chicago Press, Chicago, 1967.

---

[10]In an earlier, related essay, Bazerman explains the importance of empirical facts to modern science. ". . . [a] fact once discovered and expressed gives the scientist a solid point against which to fix an argument. The mark of modern science is its active pursuit of passive constraints, maximizing empirical experience to minimize thought caprice" [45, p. 162].

14. C. Geertz, *The Interpretation of Cultures*, Basic Books, New York, 1973.
15. P. Feyerabend, *Against Method*, Verso Books, New York, 1988.
16. P. L. Berger and T. Luckmann, *The Social Construction of Reality: A Treatise in the Sociology of Knowledge*, Doubleday, Garden City, New Jersey, 1966.
17. K. Knorr-Cetina, *The Manufacture of Knowledge: An Essay on the Constructivist and Contextual Nature of Science*, Pergamon, Oxford, England, 1981.
18. B. Latour and S. Woolgar, *Laboratory Life: The Construction of Scientific Facts*, Princeton University Press, Princeton, New Jersey, 1979.
19. K. J. Gergen, The Social Constructionist Movement in Modern Psychology, *American Psychologist, 40*:3, pp. 266-275, 1985.
20. R. Rorty, *Philosophy and the Mirror of Nature*, Princeton University Press, Princeton, New Jersey, 1979.
21. R. Rorty, *Objectivity, Relativism, and Truth*, Cambridge University Press, Cambridge, England, 1991.
22. W. V. Quine and J. S. Ullian, *The Web of Belief*, Random House, New York, 1978.
23. R. Rorty, Inquiry as Recontextualization: An Anti-Dualistic Account of Interpretation, in *The Interpretive Turn*, D. R. Hiley, J. F. Bohman, and R. Shusterman (eds.), Cornell University Press, Ithaca, New York, 1991.
24. K. A. Bruffee, Social Construction, Language, and the Authority of Knowledge: A Bibliographic Essay, *College English, 48*:8, pp. 773-790, 1986.
25. S. E. Cozzens and T. F. Gieryn (eds.), *Theories of Science in Society*, Indiana University Press, Bloomington, Indiana, 1990.
26. R. Hagendijk, Structuration Theory, Constructivism, and Scientific Change, *Theories of Science in Society*, S. E. Cozzens and R. Hagendijk (eds.), Indiana University Press, Bloomington, Indiana, 1990.
27. C. Waddell, The Role of Pathos in the Decision-Making Process: A Study in the Rhetoric of Science Policy, *Quarterly Journal of Speech, 76:*4, pp. 381-400, 1990.
28. K. B. LeFevre, *Invention as a Social Act*, Published for the Conference on College Composition and Communication, Southern Illinois University Press, Carbondale, Illinois, 1987.
29. C. Bazerman, *Shaping Written Knowledge: The Genre and Activity of the Experimental Article in Science*, University of Wisconsin Press, Madison, Wisconsin, 1988.
30. D. Journet, Writing, Rhetoric, and the Social Construction of Scientific Knowledge, *IEEE Transactions on Professional Communication, 33*:4, pp. 162-167, 1990.
31. C. R. Miller, Some Perspectives on Rhetoric, Science, and History, Revision of D. H. McCloskey, *The Rhetoric of Economics*, J. S. Nelson, A. Megill, and D. N. McCloskey (eds.), *The Rhetoric of the Human Sciences: Language and Argument in Scholarship and Public Affairs*, and C. Bazerman, *Shaping Written Knowledge: The Genre and Activity of the Experimental Article in Science, Rhetorica, VII*:1, pp. 101-114, 1989.
32. L. Odell and D. Goswami, *Writing in Nonacademic Settings*, Guilford Press, New York, 1985.
33. L. Faigley, Non-Academic Writing: The Social Perspective, in *Writing in Non-Academic Settings*, L. Odell and D. Goswami (eds.), Guilford Press, New York, 1985.

34. T. Kent, On the Very Idea of a Discourse Community, *College Composition and Communication, 42*:4, pp. 425-445, 1991.
35. M. M. Lay and W. M. Karis, *Collaborative Writing in Industry: Investigations in Theory and Practice*, Baywood, Amityville, New York, 1991.
36. L. Odell, Beyond the Text: Relations Between Writing and Social Context, *Writing in Nonacademic Settings*, L. Odell and D. Goswami (eds.), Guilford Press, New York, 1985.
37. J. Paradis, D. Dobrin, and R. Miller, Writing At Exxon ITD: Notes on the Writing Environment of an R&D Organization, in *Writing in Nonacademic Settings*, L. Odell and D. Goswami (eds.), Guilford Press, New York, 1985.
38. C. R. Miller and J. Selzer, Special Topics of Argument in Engineering Reports, in *Writing in Nonacademic Settings*, L. Odell and D. Goswami (eds.), Guilford Press, New York, 1985.
39. N. Roundy Blyler and C. Thralls (eds.), *Professional Communication: The Social Perspective*, Sage, Newbury Park, California, 1993.
40. C. Thralls and N. Roundy Blyler, The Social Perspective and Professional Communication: Diversity and Directions in Research, in *Professional Communication: The Social Perspective*, N. Roundy Blyler and C. Thralls (eds.), Sage, Newbury Park, California, 1993.
41. T. Kent, Formalism, Social Construction, and the Problem of Interpretive Authority, in *Professional Communication: The Social Perspective*, N. Roundy Blyler and C. Thralls (eds.), Sage, Newbury Park, California, 1993.
42. D. A. Winsor, Engineering Writing/Writing Engineering, *College Composition and Communication, 41*:1, pp. 58-70, 1990.
43. D. A. Winsor, The Construction of Knowledge in Organizations: Asking the Right Questions about the Challenger, *Journal of Business and Technical Communication, 6*:1, pp. 123-127, 1992.
44. P. M. Dombrowski, Challenger and the Social Contingency of Meaning: Two Lessons for the Technical Communication Classroom, *Technical Communication Quarterly, 1*:3, pp. 73-86, 1992.
45. C. Bazerman, Scientific Writing as a Social Act: A Review of the Literature of the Sociology of Science, in *New Essays in Technical and Scientific Communication: Research, Theory, and Practice*, P. V. Anderson, R. J. Brockmann, and C. R. Miller (eds.), Baywood, Amityville, New York, 1983.
46. Einstein, Albert. Relativity: The Special and the General Theory, Trans. Robert W. Lawson, Crown, New York, 1961.
47. T. Kuhn, The Natural and the Human Sciences, in *The Interpretive Turn*, D. R. Hiley, J. F. Bohman, and R. Shusterman (eds.), Cornell University Press, Ithaca, New York, 1992.
48. E. F. Keller, The Gender/Science System: or, Is Sex to Gender as Nature Is to Science? in *Feminism & Science*, N. Tuana (ed.), Indiana University Press, Bloomington, Indiana, 1989.
49. United States, Presidential Commission on the Space Shuttle Challenger Accident, *Report to the President by the Presidential Commission on the Space Shuttle Challenger Accident*, 86-16083, GPO, Washington, D.C., 1986.

50. United States, Cong. House, *Investigations of the Challenger Accident*: Report of the Committee on Science and Technology, House of Representatives, Ninety-Ninth Congress, 87-4033, GPO, Washington, D.C., 1987.
51. D. R. Hiley, J. F. Bohman, and R. Shusterman (eds.), *The Interpretive Turn: Philosophy, Science, Culture*, Cornell University Press, Ithaca, New York, 1991.
52. N. Evernden, *The Social Creation of Nature*, The Johns Hopkins University Press, Baltimore, Maryland, 1992.

# *CHALLENGER* AND THE SOCIAL CONTINGENCY OF MEANING: TWO LESSONS FOR THE TECHNICAL COMMUNICATION CLASSROOM

## *Paul M. Dombrowski*

The *Challenger* disaster was both a terrible tragedy and a vivid personal event in the life of most Americans. Any lessons drawn from this disaster for the technical communication classroom naturally provide both a tie to the emotionally powerful personal experiences of the audience and a real illustration of the grave consequences of failed human communication. The purpose of this paper is pedagogical, presenting one way I have discussed this disaster in my technical communication classes, highlighting the social contingency of meaning. By focusing on *Challenger*, the discussion of the sociology of knowledge, a topic which seems almost required in the contemporary technical communication classroom, can be moved from abstraction to vivid, personal reality. I will begin by reviewing other articles and explaining the need for a very general observation, then review critiques of positivistic attitudes toward language, and finally discuss my classroom presentation of an authoritative narrative of two crucial aspects of the disaster, emphasizing that the meaning of knowledge is contingent on social assumptions, conceptualizations, and construction.

My primary purpose is to convince those technical communication students holding positivistic preconceptions about technical communication (especially those from the sciences and engineering) of the sociology of knowledge and meaning in the hope of preventing similar disasters in the future. Such residues of positivism impede our students' appreciation of the generative, social power of their writing. These fundamentally disempowering notions, characterized by the the window pane or conduit theories of language, assume that the objects of technical communications are "out there" already, existing in some absolute sense prior to and separate from social constructions, negotiations, and interests. These misconceptions also make the epistemological mistake of assuming that the thing-in-itself can be known unequivocally. I examine the *Challenger* disaster not in its particular causes but as an illustration of a very general phenomenon: communication is not the simple reflection of a pre-existent reality but the social, creative, interested, and often unwittingly formation of meaning. In a separate article I explore related lessons drawn from the investigations of the disaster [1].

Other authors have explored the theoretical and empirical aspects of the social construction of scientific knowledge in other areas. Journet, for example, reviews several major compendiums of instances [2]. Still others have examined the particulars of this disaster. The Presidential commission and the Congressional committee present a complex history of events and causes. Winsor (1988) shows instances of interpreting from different perspectives and the difficulty of accepting "bad news" [3]. Winsor also shows (1990) how the very questions asked in investigation guide the formation of our understanding and how rhetoric can facilitate the communicative tasks of engineers and managers [4]. Pace has examined the group differentiation process as it affected the decision to launch [5]. Gouran et al. present a social psychological perspective based on the commission's finding that the procedural system itself was sound [6]. Herndl et al. present a rhetorical and argumentative perspective concluding that it is difficult to neatly define the "discourse communities" in this event [7]. Relatedly, Zappen points out the complexity and methodological contingency of "discourse community" [8].

These other authors have struggled to explain and interpret incidents relating to the disaster in order to prevent similar disasters in the future. They have identified various loci of responsibility, causality, and intervention ranging from personal judgment to hierarchical organization [3, 9]. These explorations, however valuable in themselves, collectively yield a morass of competing explanations. For example, Zappen suggests that there are several understandings of what constitutes a "discourse community" [8]. As another example, Winsor (1988) suggests the hierarchical nature of the NASA contributed to failed communication while, on the other hand, Perrow states that complex, high-risk technologies such as the Shuttle *require* an authoritarian and hierarchical organization [3, 10]. Rather than attempt to sort out this morass, for instructional purposes I focus only on the general phenomenon of the social contingency of the meaning of knowledge. The simple but vital awareness that data do not speak for themselves and that the meaning of a "fact" is contingent on many social factors rather than compelled by its own autonomous authority, is perhaps the most general and least equivocal pedagogical lesson to be drawn from the disaster.

Additionally, in the same context, I discuss in class the various articles written about the disaster, which in themselves also demonstrate the indefiniteness and the social contingency of meaning. None of these authors completely agrees either as to interpretation or as to prevention, yet they are all examining the same protracted, complex event. The data, we might say, remain constant while what they signify or mean is practically an independent variable.

Winsor (1990), a thorough, thoughtful exploration of the social construction of knowledge relating to *Challenger*, is a good illustration of the difficulty of searching for definite explanations for the disaster [4]. After reviewing various articles and then exploring in her own earlier article the difficulties in defining "sound" versus "erroneous" knowledge after-the-fact, she concludes that different initial questions are called for, yet offers no definite answer to these questions. This is not to derogate Winsor's article but only to illustrate the great difficulty attending trying to pinpoint convincingly specifc causes, responsibilities, and explanations.

Therefore, I settle in my own classes for indicating the general phenomenon of social contingency in its manifold complexity then pointing out the complex organizational, political, ethical, and rhetorical ramifications of this phenomenon. I conclude by indicating that there are no easy answers or quick, sure solutions for situations such as this, stressing that indeed it can be a profoundly serious mistake to expect definite, positive answers.

## POSITIVISM AND RHETORIC

Many theorists of technical communication have pointed out that the view of knowledge and meaning that seems to underlie both early treatments of technical communication and many of our students' naive perceptions of technical communication, is largely positivistic. Halloran and Whitburn trace the excessive concern with material objects, referentiality, and spare language to both ancient rhetoricians and early modern science [11]. Such critics point out that language never does, even in the most rigorously technical of technical communications, act as a simple window pane. Language never presents a referent without distortion, or more correctly, without interpretation. Furthermore, the priority (in the sense of already-existent) of the referent implied in such a view is, as Burke and social constructionists have pointed out, a misconception. As Burke put it poetically, ". . . And how things are / And how we say things are / Are one" [12, p. 56]. Both Halloran and Miller encourage us to be less suspicious of language than positivists have been, to see language as an inseparable ally, or at least as a neutral, rather than as an enemy [13, 14]. More importantly, they have illuminated the vital connection between language and our scientific and technological culture. Indeed, the entire topic of social constructionism, associated currently with a host including Berger and Luckmann, Geertz, Knorr-Cetina, Latour, Bazerman, Myers, and LeFevre is based on this same essentially anti-positivistic understanding of knowledge and meaning [15-21].

In the classroom, I synthesize from two key incidents attending the disaster a narrow but far-reaching general lesson for the technical communication classroom: the "objects" of our communications are oftentimes not material objects and raw data but the socially contingent meanings, interpretations, and significances attached to material objects. I do this by presenting summary histories of the development of the interpretive meaning of the charring of the O-rings and the meaning of "flightworthiness" regarding recommendations for launch. I reinforce this lesson through the fact of the multiplicity of articles on *Challenger* with various understandings of construction, explanation, and intervention. As pointed out earlier, rather than explore the specific social forces at work or weigh the relative impacts of them (a formidable and inconclusive task), I settle for demonstrating the general phenomenon of social contingency. Furthermore, rather than use "social construction" which suggests active, conscious deliberations, I use "social contingency" which includes unwitting and passive considerations as well.

## CHARRING OF THE O-RING SEALS

The first incident I discuss in class is the charring of the O-ring seals beginning several years before the disaster. I proceed by presenting a summary narrative of the history of O-ring charring and the meaning attributed to it gleaned from the evidence, testimony, and findings of both the Presidential commission and the Congressional committee [22, 23]. The basic information has not been seriously challenged though others have, not surprisingly, offered differing interpretations of this material.

I begin with a brief historical and technical overview. The Solid Rocket Booster, two of which are attached to the shuttle and its fuel tank like two enormous Fourth-of-July skyrockets, is a huge structure. It is so large that it could not be fabricated as a single structure by technology circa 1970. It had to be fabricated as several segments which were bolted together to form an entire booster. The seal between these massive segments was of vital importance. If any of the hot, explosively pressurized exhaust gases vented through the side of the booster rather than through the nozzle, the gases would immediately erode an increasingly large hole like water through a hole in a dike. The importance of the seal called for it to be redundant, meaning that a double seal had to be used—if one failed, there was the other to ensure safety. The seals were rather simple and surprisingly fragile rubber O-rings. They were separated from the hot exhaust gases by a generous glob of putty which was to perfect the sealing and protect the O-rings from being burned and made inoperative. That is, neither O-ring was expected to ever come in contact with hot exhaust gases because such exposure would immediately threaten the loss of the seal, the integrity of the booster, and so the entire vehicle and crew.

The boosters, due to their high initial cost and basic simplicity, were to be re-used over the course of several Shuttle missions. After completing their pyrotechnic boost of the main vehicle, they were disengaged at high altitude and parachuted to fall in the ocean. They were then recovered and recharged for another flight.

From the earliest Shuttle missions, however, it was apparent that something was wrong. Examination of the spent boosters revealed that the O-rings frequently were charred to varying degrees, some almost half-way through. Keep in mind that the O-rings were intended never to be exposed to exhaust gases in any way at any time. Thus, this charring was at first said to be "anomalous," that is, it was not supposed to happen.

The earliest observations of charring were noted with alarm and reported to higher authorities because the seals were not operating as expected and because sound sealing was vital to the safety of the flight. For a variety of reasons beyond the scope of this paper, these alarms were noted but little was done to change the design of the seal system or to curtail flights until the problem was corrected. Instead, procedures were instituted to apply more putty, install the O-rings more carefully, and test more scrupulously the seating of the O-rings. To the credit of Morton Thiokol Industries (MTI) and the National Aeronautics and Space Administration (NASA), some organizational steps were taken to begin a long-range, more substantial remedy. However, as investigating committee documents clearly reveal, this task force was "hog-tied by paperwork" and continually "delayed" [23, p. 57; 22, pp. 252, 253]. Its efforts came to little as increasingly vocal and urgent warnings apparently went unheeded or were reconceptualized and dismissed.

The instructive aspect of this charring episode is that social factors (economic, historical, political, professional, organizational, and rhetorical concerns) had a powerful effect on how this anomaly was perceived and what was to be done in light of that perception. Indeed, whether the charring was construed as "anomalous" or not was socially determined, as we will see, this despite seemingly obvious indicators of danger.

As flight after flight was launched and successfully recovered even though some charring of some O-rings occurred, these flights were taken as a sort of evidence that charring should be understood in a new light. The very success of these flights was taken to demonstrate that exposure of the O-rings to the exhaust gases was not a serious concern and could and should be tolerated. The Congressional report points out seven instances of "poor technical decisions" leading to the disaster, one of which is "Mr. Mulloy's description of joint failures as being within their 'experience base.' In other words, if it broke before and the size of the recent break was no bigger than those before, then there was no problem. Even when the erosion surpassed all previous experience, NASA then went on and expanded its 'experience base'" [23, p. 50]. That is, the successful completion of the mission was taken as *prima facie* evidence that exposure and charring should be tolerated. Thus what was never to happen came to be permissible, even being taken as an indication of safety rather than danger.

The Congressional committee report is especially enlightening on this matter, tracing the introduction of "acceptable erosion," "allowable erosion," and "acceptable risk" into discussions of the charring [23, p. 53, 55, 56]. "But rather than identify this condition as a joint that didn't seal, that is, a joint that had already failed, NASA elected to regard a certain degree of erosion or blow-by as 'acceptable'" [23, p. 62]. As time went on, therefore, the increasing number of unanticipated events came to be viewed with decreasing concern. Thus, the anomalous was no longer considered anomalous because it happened all the time, and what was cause for alarm became grounds for reassurance. It was as though black became white.

The Congressional committee is also clear in finding that the information available before launch was not equivocal and should have prevented launch. "The joint seal problem was recognized by engineers in both NASA and Morton Thiokol in sufficient time to have been corrected by redesigning and manufacturing new joints before the accident on January 28, 1986 [23, p. 50]. More pointedly, "The question remains: Should the engineering concerns, as expressed in the pre-laucnh teleconference [the L-1 conference the night before the launch in which flightworthiness was discussed in relation to seal erosion history], been sufficient to stop the launch? The Committee concludes that the answer is yes" [23, p. 71].

The lesson to be drawn from this reversal of meaning is that data are not the clear, absolute, pre-existent entities that a positivistic pre-conception would suggest. Instead the meaning of data is understood and defined in light of many other social considerations, some of which are necessary and appropriate (such as the replicability of the scientific method) and some of which clearly are neither. Expecting the data to speak for themselves, to tell their own story as though from their own autonomous authority, is to disregard the social contingency of knowledge and meaning.

## FLIGHTWORTHINESS

The second episode has to do with the teleconference the night before the launch (called the L-1 meeting) between NASA managers and MTI managers and engineers. During this teleconference, management and engineers at MTI strenuously expressed their grave reservations about the safety of the flight on the basis of the charring of the O-ring seals (some engineers refusing to buy into the reconceptualization of the anomaly). The engineering group at MTI was "very adamant about their concerns . . . because we were way below our data base and we were way below what we qualified for" [22, pp. 86-89]. The response of management to this expression of reservations at such a late date was to say that they were "appalled" (Hardy at Marshall SFC) because such reservations called up the possibility of postponing the already long-postponed flight for another three months at least. MTI management acted by calling for an off-line caucus in which they discussed among themselves the seriousness of these reports, the implications, the responsibility and authority involved, and the actions to be taken. Management (Mason) told engineers that what was needed was a "management" decision rather than an engineering decision, which ultimately had the effect of overriding the objections of the engineers.

Management at MTI and NASA questioned engineers further, pressing them to prove that their reservations involved certain peril to the mission. Thus began the second major reconceptualization. Management's questioning (which could be construed as brow-beating, considering the power and status differential) expressed a complete change in perspective and assumption. Engineers found themselves in the situation of being asked by management to prove absolutely and certainly that the *Challenger* flight would end in disaster. This they could not do, especially in light of previous successful missions in which some charring had occurred. More specifically, engineers were totally thrown into confusion and frustration by the change of assumption and conceptualization. The standard perspective for these discussions with management was that engineers were called on to prove that the vehicle *would* fly safely, working to refute the sort of devil's advocate approach by management which assumed that it *would not* fly. This standard assumption prudently leaves the engineers with the burden of proof; without convincing proof, the flight is scrubbed.

At the *Challenger* L-1 meeting, however, there occurred an absolutely vital flip-flop of assumption and burden of proof. When engineers could not prove that *Challenger would not* certainly go up in flames, management took their implicit assumption that *Challenger* would fly as thus confirmed and unqualified.

Lund testified to the Presidential commission:

> We . . . have always been in the position of defending our position to make sure that we were ready to fly . . . . I didn't realize until after that meeting and after several days that we had absolutely changed our position from what we had been before . . . . We had to prove to them that we weren't ready, and so we got ourselves in the thought process that we were trying to find some way to prove to them that it wouldn't work, and we were unable to do that. We couldn't prove absolutely that that motor wouldn't work . . . . It seems like we have always been in the opposite mode [22, p. 94].

Boisjoly, an engineer at MTI, later summed up the alteration of assumptions succinctly:

> This was a meeting where the determination was to launch, and it was up to us to prove beyond a shadow of a doubt that it was not safe to do so. This is in total reverse to what the position usually is in a preflight conversation or flight readiness review. It is usually exactly opposite to that [22, p. 93].

Nevertheless, and to their credit, MTI engineers continued to vociferously object to approving the launch. But in the face of an adamant management and an argumentative perspective that was not only unusual but which opposed accepted good practice in astronautical engineering, the engineers gave up. As Boisjoly testified to the Presidential commission:

> And we were attempting to go back and re-review and try to make clear what we were trying to get across, and we couldn't understand why it was going to be reversed. So we spoke out and tried to explain once again the effects of low temperature. Arnie [Thompson] . . . tried to sketch out once again what his concern

> was with the joint, and when he realized he wasn't getting through, he just stopped. I tried one more time with the photos . . . . I also stopped when it was apparent that I couldn't get anybody to listen [22, p. 92].

In this case, the all-important meaning to be attached to an event (the engineers' unwillingness to recommend the launch and the charring history itself) was pre-determined by NASA management. Further objecting by MTI engineers did not fit this prior conceptualization and so was discounted by management or not even recognized: "nobody said a word" [22, p. 92].

Bringing the discussion of this episode to a close, I point out the interrelation of the interpretation of O-ring erosion history and the reversal of the assumption of unflightworthiness. I quote from the Presidential report where Boisjoly is queried by Feynman about the L-1 conference regarding his inability to *prove* unflightworthiness.

> Feynman: I take it you were trying to find proof that the seal would fail?
>
> Boisjoly: Yes.
>
> Feynman: And of course, you didn't, you couldn't, because five of them didn't, and if you had proved that they would have all failed, you would have found   yourself incorrect because five of them didn't fail.
>
> Boisjoly: That is right. [22, p. 93].

When the assumption was changed to having to prove that *Challenger* would *not* fly, engineers could not prove this with certainty because five earler flights had returned safely with charring. This reinforces the crucial importance of the conceptualization of O-ring erosion.

This episode shows that prior conceptualizations can work both to alter drastically the meaning of evidence and even to refuse to recognize conflicting evidence. The crucial question of whether the shuttle was flightworthy or not at a given time was ultimately answered less by hard data or even by the interpretation advanced by engineers but more by the conceptualization advanced by another, more powerful social group.

## LESSONS FOR TECHNICAL COMMUNICATION

What do these two episodes suggest for the technical communication classroom? They raise the question, what is technical communicators about? Is it, as some students with positivistic preconceptions believe, only about objects and data, the artifacts of technology? The lesson of these two episodes answers, No. It is clear that the object, in the case of the O-ring and its physical condition, was not in itself the crucial factor in the communications. That charring occurred is indisputable but also, in a way, trivial. Rather, what is important and problematic and what gave purpose to communications was the interpretation or meaning of this charring and what should be done in light of it. Does the charring mean we should postpone the launch or should we not? In this case, the attendant assumptions and interests were absolutely vital to defining the substantive content and purpose of technical communication about the charring.

In the case of the L-1 meeting, the social construction imposed on the interaction by management (from their position of greater authority and power) completely undid the conventional assumptions of engineers regarding good engineering practice. The reversal of argumentative assumption in effect reinterpreted the positivistic data about the charring by casting it in completely different light, yielding conclusions unexpected by engineers. Perhaps equally importantly, this extraordinary conceptual reversal undid any force to the engineers' data to the point that there was nothing possible for them to say. It also undid the lives of seven people.

The lesson of the *Challenger* disaster are simple, powerful yet at the same time difficult to grasp fully. The thing, the charred seal, is itself of little interest or even meaning in itself for technical communication. Only when we begin to grapple with such questions as, what does this mean, how is it to be interpreted, and what are we to do in light of this, does a valid rhetorical object emerge for technical communication. This is true for practically all technical reports in that a purpose statement casts the context reports in light of on-going social, organizational, professional, and discourse community concerns while the summary section and conclusions and recommendations section cast the raw material of the report in interpretations and applications, i.e., constructions, along the line of social interests broadly defined.

## OTHERS ON
## SOCIAL CONSTRUCTION OF MEANING

Karen Burke LeFevre on writing is particularly germane regarding the social contingency of meaning. LeFevre explains that invention in writing is an inherently social act, an essentially "communitarian" endeavor. The sociality of invention stands in opposition to some prevailing assumptions about writing which seem particularly prevalent in the technical communication classroom. Among these are the assumption is that "knowledge is represented in language rather than constituted by it" and that "'hard' knowledge is mechanistically gained by the accumulation of objective facts" [21, p. 135]. As the *Challenger* examples show, knowledge is instead constituted in such communication acts as the reconceptualization of assumptions and accumulated evidence. In this case the increasing accumulation of evidence was counterproductive of safety because of the reconceptualization of "anomalous" or the reconceptualization of "flightworthiness" imposed on the evidence as an *a priori*. LaFevre argues that

> We should seek . . . to persuade those who write about science and technology—and their employers—that writing and language are closely connected with invention or innovation, that they do much more than merely transmit work that has already been completed [21, p. 135].

The epistemological indeterminacy of such "things" as the charring of an O-ring should not be surprising. Paul Feyerabend and others interested in the sociology of science have pointed out that all knowledge, including scientific and technological knowledge, is sociologically contingent, being based on a tradition or context of consensus, semantics, and even cognition. Feyerabend emphasizes that even in the most supposedly objectivist science, physics, knowledge is socially contingent in many ways. He explains that the assumption of fundamental indeterminacy and the toleration of alternative construals are the *sine qua non* of a growing, vibrant science. "Science is an essentially anarchic enterprise: theoretical anarchism is more humanitarian and more likely to encourage progress than its law-and-order alternative" (which will tolerate only a single interpretation) [24, p. 5]. This amounts to saying that the epistemological condition in which the O-ring charring could be construed either as a cause for alarm or as a cause for assurance is fundamentally inescapable and a basic feature of the advancement of knowledge. This is not to say, however, that the inclination to make safety the paramount concern should not be operative in the minds of those construing the "thing."

David Dobrin holds similar views. Dobrin points out the vital importance of alternity in human communication, explaining that the degree of alternity and flexibility is in a way a distinction between literary expressions and technical expressions (at least those from the window pane perspective). He also points out that too often the technical communicator feels "dominated" and "subjugated" in the face of technology [25, p. 245]. He argues that instead the communicator, by virtue of knowing so much more than most readers about what he or she writes, actually makes *de facto* policy and thus actually has substantial power. He also explains that the communicator stands in a unique relation, straddling the otherwise separated groups of technicians and users, and so performs a valuable social function.

Carolyn Miller critiques the window pane view of language in technology and science as being suspicious of language while at the same time fostering a socially invidious impersonalness. When science and technology insist on privileging atomistic knowledge about content and objects, it disempowers those not participating in that privilege. Miller argues instead that "reality cannot be separated from our knowledge of it; knowledge cannot be separated from the knower; the knower cannot be separated from a community. . . .Science, then, is not concerned directly with material things, but with these human constructions, with symbols and arguments" [14, pp. 615-616]. Miller advocates a humanistic education for technical communicators. "To write, to engage in any communication, is to participate in a community; to write well is to understand the conditions of one's own participation—the concepts, values, traditions, and style which permit identification with that community and determine the success or failure of communication" [14, p. 616]. This communitarian awareness includes recognition of the fundamental interpretative indeterminacy and social contingency of a great deal of technical knowledge.

LeFevre, Feyerabend, Dobrin, and Miller all agree that the positivistic, window pane assumption that the objective, material world is absolute and amenable to unisignatory expression in an ideally referential way, is fundamentally invalid. It is therefore incumbent on teachers of technical communication to point out both the socially contingency of the meaning of "things" and the social responsibility which this contingency entails.

## CONCLUSION

The two aspects of the *Challenger* disaster which I have discussed vividly demonstrate the powerful social contingency of meaning. The raw evidence, having little meaning in itself, has importance contingent on social assumptions, conceptualizations, and constructions. In the first instance, what is meant by "anomalous" and what are the conditions under which anomalousness is to be attached to an observation was instrumental in reconceptualizing a grave danger into an assurance of safety. In the second instance, the subtle reconceptualization of the assumption regarding flightworthiness, do we assume that *Challenger* will fly or do we assume that it will not, was crucial in defining the light in which engineers' data were seen, in turn directing the course of the decision to launch or not. The outcome we all know.

## REFERENCES

1. P. M. Dombrowski, The Lessons of the *Challenger Disaster, IEEE Transactions on Professional Communication, 34*:4, pp. 211-261, 1991.
2. D. Journet, Writing, Rhetoric, and the Social Construction of Scientific Knowledge, *IEEE Transactions on Professional Communication, 33*, pp. 162-167, 1990.
3. D. A. Winsor, Communication Failures Contributing to the Challenger Accident, *IEEE Transactions on Professional Communication, 31*:3, pp. 101-107, 1988.
4. D. A. Winsor, The Construction of Knowledge in Organizations: Asking the Right Questions about the Challenger, *Journal of Business and Technical Communication, 4*:2, pp. 7-20, 1990.
5. R. C. Pace, Technical Communication, Group Differentiation, and the Decision to Launch the Space Shuttle Challenger, *Journal of Technical Writing and Communication, 18*:3, pp. 207-220, 1988.

6. D. S. Gouran, R. Y. Hirokawa, and A. E. Martz, A Critical Analysis of Factors Related to Decisional Processes Involved in the Challenger Disaster, *Central States Speech Journal, 37*:3, pp. 119-135, 1986.

7. C. G. Herndl, B. A. Fennell, and C. R. Miller, Understanding Failures in Organizational Discourse: The Accidents at Three Mile Island and the Shuttle Challenger, in *Textual Dynamics of the Professions: Historical and Contemporary Studies in Writing in Professional Communities,* University of Wisconsin Press, Madison, Wisconsin, 1991.

8. J. P. Zappen, The Discourse Community in Scientific and Technical Communication: Institutional and Social Views, *Journal of Technical Writing and Communication, 19*:1, pp. 1-11, 1991.

9. P. M. Dombrowski, *People or Procedures?: Ethical Lessons of the Challenger Disaster*, paper presented at 1989 Conference of The Society for the Social Studies of Science, Irvine, California, 1989.

10. C. Perrow, *Normal Accidents: Living with High-Risk Technologies*, Basic Books, New York, 1984.

11. S. M. Halloran M. Whitburn, Ciceronian Rhetoric and the Rise of Science: The Plain Style Reconsidered, in *The Rhetorical Tradition and Modern Writing*, J. J. Murphy (ed.), Modern Language Association, New York, 1982.

12. K. Burke, *Language as Symbolic Action*, University of California Press, Berkeley, California, 1966.

13. S. M. Halloran, Eloquence in a Technological Society, *Central States Speech Journal, 29*, pp. 221-227, 1978.

14. C. Miller, A Humanistic Rationale for Technical Writing, *College English, 40*:6, pp. 610-617, 1979.

15. P. L. Berger and T. Luckmann, *The Social Construction of Reality: a Treatise in the Sociology of Knowledge*, Doubleday, Garden City, New York, 1966.

16. C. Geertz, *Works and Lives: The Anthropologist as Author*, Stanford University Press, Stanford, California, 1988.

17. K. Knorr-Cetina, *The Manufacture of Knowledge: An Essay on the Constructivist and Contextual Nature of Science*, Pergamon, Oxford, United Kingdom, 1981.

18. B. Latour, *Science in Action: How to Follow Scientists and Engineers through Society*, Harvard University Press, Cambridge, Massachusetts, 1987.

19. C. Bazerman, *Shaping Written Knowledge: The Genre and Activity of the Experimental Article in Science*, University of Wisconsin Press, Madison, Wisconsin, 1988.

20. G. Myers, *Writing Biology: Texts in the Social Construction of Scientific Knowledge*, University of Wisconsin Press, Madison, Wisconsin, 1990.

21. K. Burke LeFevre, *Invention as a Social Act, Published for the Conference on College Composition and Communication*, Southern Illinois University Press, Carbondale, Illinois, 1987.

22. United States, Presidential Commission on the Space Shuttle Challenger Accident, *Report to the President by the Presidential Commission on the Space Shuttle Challenger Accident*, 86-16083, GPO, Washington, D. C., 1986.

23. United States, Congress House, *Investigation of the Challenger Accident: Report of the Committee on Science and Technology*, House of Representatives, Ninety-ninth Congress, 87-4033, GPO, Washington, D. C., 1987.

24. P. Feyerabend, *Against Method*, Verso, New York, 1990.

25. D. Dobrin, What's Technical About Technical Writing, in *New Essays in Technical and Scientific Communication: Research, Theory, Practice*, P. V. Anderson, R. J. Brockmann, and C. R. Miller (eds.), Baywood, Amityville, New York, 1983.

# SOME PERSPECTIVES ON RHETORIC, SCIENCE, AND HISTORY

## Carolyn R. Miller

### A REVIEW ARTICLE OF

*The Rhetoric of Economics,* by D. N. McCloskey. Madison: University of Wisconsin Press, 1985. pp. xx + 209.
*The Rhetoric of the Human Sciences: Language and Argument in Scholarship and Public Affairs,* ed. J. S. Nelson, A. Megill, and D. N. McCloskey. Madison: University of Wisconsin Press, 1987. pp. xiii + 445.
*Shaping Written Knowledge: The Genre and Activity of the Experimental Article in Science,* by C. Bazerman. Madison: University of Wisconsin Press, 1988. pp. xi + 356.

A footnote in a recent issue of this journal (*Rhetorica*) provides a provocative context for a discussion of books in the University of Wisconsin's series on the Rhetoric of the Human Sciences. In the footnote, Brian Vickers cites an advertisement for the series as evidence that the term "rhetoric" is used "quite indiscriminately" in book titles and publishers' announcements: the advertisement says that the series will analyze "various disciplines, not as the 'sciences' they sometimes claim to be, but as 'rhetorics'—that is, as systems of belief and practice, each of which has its own characteristic form and structure" [1, p. 26]. Vickers' comment is that this use "widen[s] the meaning of rhetoric to the point of no return" [1, p. 26]. It is odd that this warning about what might be seen as the *hypertrophy* of rhetoric appears in an essay entitled *The Atrophy of Modern Rhetoric.*

There are at least two types of arguments against expanding rhetoric as the Wisconsin series proposes, and both are related to issues in the history of rhetoric. First is an essentialist argument, or argument from definition. One may hold that rhetorical discourse and scientific discourse are fundamentally different in kind, so different that they cannot be described and studied with the same set of concepts. This view has roots deep in the classical tradition, most firmly in Aristotle's distinction between rhetoric and analytic, between discourse designed to address the general public on matters of opinion and discourse comprehensible to the specialist on matters of natural necessity. This view became more important in modernist thinking, which not only distinguished science from rhetoric but elevated science over rhetoric. (Indeed, Vickers' essay documents this configuration well and aptly characterizes it as the atrophy of rhetoric.) The second type of argument is from circumstance. One may hold that, since rhetoric is a particular theory of discourse, with particular historical sources, the term is best reserved for uses entirely consistent with that history so as not to misread the past or confuse the present. If changing historical conditions lead to changes in discourse theory, one had best find another terminology that signals the change; for example, instead of using rhetoric to describe "systems of belief and practice," one might use the term "ideology."

A great deal of twentieth-century, "postmodern" thinking has challenged the essentialist argument, primarily by denying that science can be what modernism requires: a methodological skepticism that leaves indubitable truth as its residue. Science is seen to be, in fact, rather like rhetoric—it requires the persuasion of an audience, it proceeds by argumentation, it uses topics and tropes, it relies on premises that are never fully demonstrable, it produces knowledge that can never be certain or necessary, it exists in discourse. Once one concedes the essentialist position, the historical particularity of rhetoric is harder to maintain, so the circumstantial position is also vulnerable. But to attack these arguments separately, one might begin by noting that even within the classical and modernist traditions, "rhetoric" has already been used to cover a multitude of discourse theories. In addition, one can look for evidence that the traditional vocabulary of rhetoric can be usefully applied to the discourse of science and for other advantages of conceiving of rhetoric in expanded or historically adapted senses. Richard McKeon envisioned such advantages. He once caricatured the purist approach to rhetorical history as "the monotonous enumeration of doctrines, or preferably sentences, repeated from Cicero or commentators on Cicero" [2, p. 120]. Instead, for McKeon, the history of rhetoric should be "the history of a continuing art undergoing revolutionary change" [2, p. 22].

The Wisconsin series on the rhetoric of the human sciences provides material for exploring these issues concerning the relationship between science, rhetoric, and versions of history. The three Wisconsin books to be considered here, of the dozen or so now in print and the two dozen projected volumes, are Donald McCloskey's *The Rhetoric of Economics* (1985), Charles Bazerman's *Shaping Written Knowledge* (1988), and Nelson, Megill, McCloskey's collection on *The Rhetoric of the Human Sciences* (1987). The first two are the books so far available that most closely concern the "harder" sciences, economics and experimental physics being more quantitative and empirical in general than law, politics, anthropology and psychotherapy, the subjects of other books in the series. The third book, an essay collection from a 1984 symposium at the University of Iowa, covers a wide gamut of disciplines, from mathematics, paleontology, and behaviorist psychology to history, theology, and women's studies.

As a group these books accept the postmodern critique of science and promote the postmodern expansion of rhetoric, although without much advance in theoretical terms. They do add a great deal to the circumstantial evidence favoring the application of rhetoric to scientific discourse by showing the possibility and productivity of such analysis. The overall effect of the three, then, is support for the theoretical position that has come to call rhetoric "epistemic," after Robert L. Scott's 1967 essay. This position (at least in its stronger forms) denies the classical and modernist distinctions between rhetorical and scientific discourse and finds some pleasure in extending the use of the term rhetoric into new areas like science, not only to make rhetoric more powerful but also to emphasize the similarities between public and specialized discourse. It holds that knowledge, social relations, and the self are *constituted by* discourse, not just expressed or transmitted through discourse. In his review of rhetorical theory covering 1976 and 1977, Michael C. Leff points out that this position constitutes a "radical break" with the modernist tradition, although he notes that there is historical precedent for the epistemic view in the rhetoric of Gorgias and that this foregrounding of sophistic rhetoric is itself a departure in rhetorical history [3].

Donald McCloskey's book on rhetoric in the discipline of economics offers some argument against both the essentialist and circumstantial positions [4]. The book as a whole is a tract against what McCloskey considers modernist dogma in economics; it also offers ample evidence that the traditional methods of rhetorical analysis can be effectively deployed against the discourse of academic economists. To take the latter point first, McCloskey claims that "economics is a collection of literary forms, not a science" [4, p. 55]. This position invites a certain kind of analysis of those forms, analysis that McCloskey variously calls "literary" and "rhetorical." Here, he proves himself a skillful critic. He examines the writing of what a noneconomist must assume to be big names in economics: Paul Samuelson, Gary Becker, Robert Solow, John Muth, Robert Fogel. He finds metaphors, common topics, appeals to authority, special topics, appeals to ethos, metonymies, synechdoches, and ironies—all at work to persuade an audience of matters that cannot be (or at least have not been) established with certainty. He finds arguments that fall far short of the completeness and rigor required by the model of a nonrhetorical scientific discourse. He explains why a paper that seems, on the surface, "badly written" (even to economists) achieved its influence; he explains why one book was more influential than another that made the same argument at about the same time [4, p. 88]. McCloskey's detailed criticism shows that the conceptual vocabulary of traditional rhetoric can indeed elucidate the workings of discourse that would not be called rhetorical by some traditionalists. This is not conclusive proof that rhetoric should be applied to scientific discourse, but it is a strong argument from utility.

In showing that economics as a discipline falls short of the ideals of scientific methodology it sets for itself, McCloskey contributes to his other major agenda, which is to discredit modernism as a model for the discipline. Not only is economics a collection of literary forms rather than a science, so, he claims, is any science [4, 55]. In arguing against the methodology of positivism, McCloskey rehearses what are by now, after Thomas Kuhn, familiar anti-modernist points about the impossibility of necessary and certain proof. He relies also on the anti-methodism of Paul Feyerabend and the anti-epistemologism of Richard Rorty. Since McCloskey conceives of his audience as primarily economists, not philosophers or even rhetoricians, one assumes that modernist-methodologism in the field of economics is so strong that the litany bears repeating, and if his representation of the state of economic scholarship is close to accurate, this assumption seems warranted. The reviews of his book in economics journals, however, reveal a more divided opinion: of the six reviews I found, one recommended caution on modernist grounds, two accepted McCloskey's criticisms, and three urged a more thoroughgoing postmodern critique.

Where McCloskey stops short is at the crux of the epistemic hypothesis. His book, he says, is about scholarship, about the conversation of economists [4, p. xviii]. What he hopes for is that economic discourse might become more self-aware, more modest, more tolerant; that better prose, teaching, and relations with other disciplines might be possible. Rhetoric, he says, is "good for you" (and "anti-modernism is nice"), but rhetoric "will not precipitate any revolution in the substance of economics" [4, pp. 51, 50, 174]. Rhetoric is not the source of or the cause of knowledge; it is merely the means of expressing and disseminating knowledge. If economists realize and admit that they need rhetoric, they will behave better to each other, perhaps, but they will not understand their subject differently. They will continue to treat persons as "human capital," actions and decisions as motivated by simple calculations of "expected utility," social conditions as "demand curves", rationality as a calculus, as long as they can decently persuade other economists to do so [4, pp. 77, 81]. They will still understand the knowledge both of economic actors and of academic economists as solipsistic rather than social [4, p. 99]. McCloskey's understanding of rhetoric, although he has learned much from contemporary rhetoricians such as Wayne Booth, Chaim Perelman, and Kenneth Burke, is very much a literary one, one that has more to say about the forms of texts than about the substance of social belief.

A rhetoric of substance for economics is advanced by Arjo Klamer in his essay for Nelson, Megill, and McCloskey's Rhetoric of *Rhetoric of the Human Sciences* [5]. Klamer says that McCloskey "is critical of the ways in which economists talk about what they do but not of the ways in which they talk about the economy" [5, p. 164]. McCloskey noted that economists rely on unstated assumptions, which they sometimes shift in mid-argument; Klamer is concerned with what those assumptions are and how they affect the substance of the discipline and the economy itself, given the social power of economists as policy makers. The example that Klamer focuses on is the "rationality assumption," the unargued axiom that economic actors behave so as to optimize their own economic interests. He finds that although it is limited and unrealistic the assumption is justified by a network of other beliefs that form what Kuhn might call the "disciplinary matrix" of modernist economics: the supremacy of Reason, *de gustibus non est disputandum,* radical individualism, utilitarianism, mechanistic models of human behavior, as well as bureaucratic social control and the class structure of capitalist society [5, pp. 176-177, 180]. Klamer suggests that there are other versions of economic theory that do not require modernist assumptions, and although he does not describe them in detail, his analysis of the metaphors and premises of this one version shows by indirection how one could have a substantively different discipline.

The collection of twenty-two essays in which Klamer's appears brings together a great many different approaches and agendas. As the editors explain in the preface, the book is the result of a 1984 Humanities Symposium on the Rhetoric of the Human Sciences held at the University of Iowa. The symposium itself grew out of an interdisciplinary faculty seminar at Iowa that began in 1980 and involved the three editors of the collection: the economist McCloskey, political scientist Nelson, and historian Megill. The editors are a major presence in the book, represented by an introductory essay by all three, an essay on political science by Nelson, an essay on history by Megill and McCloskey, and a closing essay by Nelson. Other reflections on the symposium have been offered in *The Quarterly Journal of Speech* by John Lyne and Herbert W. Simons.

As with McCloskey's book, the major focus of the seminar, the symposium, and this collection of essays is a perspective on the discourse of scholarship, which came to be called at Iowa "rhetoric of inquiry" (characteristically without the definite article) [6]. Rhetoric of inquiry is a transdisciplinary perspective that understands all scholarly communication as rhetorical; it is, as the editors say in their introduction, "comparative epistemology" [6, p. 17]. It is contrasted with "logic of inquiry," which is the universal method promoted by scientific modernism. As in McCloskey's book, conversation is the preferred model for scholarly communication here, a model made influential by the philosopher Richard Rorty. Rorty's essay in this collection, "Science as Solidarity," urges that the goal of inquiry is "the attainment of an appropriate mixture of unforced agreement with tolerant disagreement (where what counts as appropriate is determined, within that sphere, by trial and error)." This pragmatist program, he says, would "gradually make unintelligible the subject-object model of inquiry, the child-parent model of moral obligation, and the correspondence theory of truth" [7, p. 48]. Rorty's program does not privilege scholarly inquiry in particular, or science as the model of inquiry, as traditional epistemology did; rather, he says, "the only sense in which science is exemplary is that it is a model of human solidarity"—in other words, it has come closer than any other enterprise to that appropriate mixture of unforced agreement and tolerant disagreement [7, p. 46].

If Rorty's essay makes an extreme case against the essentialist position, perhaps the strongest case against the circumstantial position is offered by the essay on *Rhetoric and Mathematics,* by Philip J. Davis and Reuben Hersh [8].

Mathematics has always seemed to be the obvious example of universal method leading to guaranteed knowledge, or, as Davis and Hersh put it, "mathematics appears as the dragon which must be slain" [8, p. 54]. *A fortiori,* if mathematics is rhetorical, the rest of the disciplines must also fall into the rhetorical parade. Davis and Hersh demonstrate that mathematics is as much rhetorical as it is logical: "A mathematical paper does two things. It testifies that the author has convinced herself and a circle of friends that certain 'results' are true. And it presents a part of the evidence on which this conviction is based" [8, p. 61]. The authors go on to emphasize two rhetorical aspects to this description—the dependence on a particular audience for criteria of "conviction," and the necessarily partial and often metaphorical nature of the proof. And they claim that there is no transcendental notion of a sufficient mathematical proof; it depends on the practices and issues in a given branch of mathematics (and, one would suppose, on its historical development). They conclude by claiming that the confidence that mathematicians have in their results "is not absolute, nor is it fundamentally different in kind from our confidence in our judgments of the realities of ordinary daily life" [8, p. 68].

Since rhetoric of inquiry is supposed to "encompass the interdependence of inquiry and communication" it would seem able to include disciplines outside the "human sciences," like mathematics, and the biological and physical sciences [6, p. 16]. But only one other essay here addresses such a field, John Campbell's on Charles Darwin as rhetorician of science. Interestingly, since Campbell is not an evolutionary biologist but a rhetorical critic, this essay relies on a conventional, modernist notion of rhetoric to explain how Darwin could be persuasive both to the general Victorian public and to a "key minority of his professional peers"; the argument relies primarily on discrepancies between Darwin's notebooks and his public accounts of his work [9, p. 69]. Campbell shows that *The Origin of Species* was "premeditated", constructed deliberately and skillfully to meet the conflicting needs of his two audiences [9, p. 77]. This essay is really not about "rhetoric of inquiry" in the epistemic sense but about the rhetoric of dissemination, about persuading others.

Although the human sciences are not a necessary focus of rhetoric of inquiry, this focus is justified for this volume at several points. According to Nelson, for example, the social sciences have the most to gain from rhetoric of inquiry, since they have been so misguided by modernist logic of inquiry, more than have the humanities and the natural sciences [10]. Charles Bazerman's essay on the development of the American Psychological Association's style manual in the context of the discipline's growing commitment to behaviorism is a specific, and ingenious, case in point. The more important justification is the connection between the human sciences and the public sphere, between, as Nelson puts it, "the academy and the polity", a connection of particular importance to rhetoric [10, p. 409]. The human sciences, it is generally assumed here, have more, or more direct, relevance to public life than other sciences; through their applications in law, policy analysis, social services, religious practice, economic activity, the human sciences bring ethics and politics into public life. One might argue with this perspective by pointing to any number of current problems with clear bases in the biological and physical sciences that also have crucial ethical and political dimensions: AIDS, the space program, computer control of stock markets and weapons systems, ozone depletion, agricultural productivity, and so forth.

The book includes essays on anthropology (by Rosaldo), paleoanthropology (Landau), psychology (Carlston), political science (Nelson), history (Megill and McCloskey), theology (Klemm), law (White), women's studies (Elshtain), and the social sciences generally (Anderson, Shapiro). These contribute to rhetorical studies mainly through their collective demonstration that, as the editors say, argument is both more unified and more diverse than usually understood [6, p. 4]. Oddly, and disappointingly, the two essays by literary scholars (Bruns, Hernadi) do not concern the discourse of their own discipline, which is surely one of the human sciences, but rather provide critical perspectives on the discourse of *other* disciplines-in-general. And because no reader is likely to be familiar with all the contributors and their diverse backgrounds, one wishes for some identification of the authors, with their affiliations, disciplines, and perhaps major accomplishments; the lack of such information is a particularly glaring editorial oversight in a book like this. What a rhetorician doesn't get from this book, by its very nature, is a sustained and consistent discussion of the rhetorics of these various fields of inquiry. The rhetorical scholar will miss the informed use of the vocabulary of rhetorical criticism, which could, potentially, provide continuity. But few rhetoricians are well enough informed about other disciplines to be able to undertake the kinds of discipline-specific critique attempted here.

Whether criticism of discipline-specific discourse is best undertaken by a disciplinary specialist or by a rhetorician is a question without an obvious answer, as John Lyne notes [11]. The relationship between rhetoric and subject matter continues to worry us long after *Gorgias*. As the editors put the issue, rhetoric seems both unified and diverse, both an art of persuasion with its own integrity and a subject-dependent, situation-specific art. In fact, one of Nelson's themes is that the modernist error was to separate method from concrete contexts, in pursuit of a logic of inquiry that would be universal [6, p. 204]. (One would thus expect there to be many rhetorics of inquiry.) Aristotle faced this same issue in attempting to establish rhetoric as a distinct and systematic art in the face of those who treated it as part of political activity. His solution consisted of distinguishing the starting points of highly specialized discourse from those most appropriate to rhetoric, the common topics, with a third kind, the specific topics, floating in between. The specific topics are subject-specific, but not very subject-specific; the more specific they get, the farther away from rhetoric you are. I have suggested elsewhere that this solution was unstable because Aristotle was not really able to differentiate between disciplinary principles and the specific topics. The distinction between rhetoric and other disciplines thus remained uneasy.

The rhetoric of inquiry project requires treating the discourse of scholarship as an analogue or extension of public discourse. Two essays demur from this project, both of them written by rhetoricians, those who by training have valued the discourse of public affairs over that of scholarship. They begin with similar doubts but reach quite different conclusions. The essay by McGee and Lyne relies on the opposition between the public and technical realms, which should remain, they believe, in "dialectical tension" [12, p. 389]. Not to preserve the distinction is to endanger the public interest and public values that rhetoric at its best represents. To use rhetoric to unify the diverse discourses of disciplines and public affairs is to endanger rhetoric itself by inviting its own ultimate methodization. Further, the conversation model that is promoted by the rhetoric of inquiry is, they believe, an inappropriate one for public discourse, since it does not recognize the exigences of power in that realm: "the political culture . . . and the culture of critical discourse stand in stark contrast to one another" [12, p. 400]. Ultimately, rhetoric cannot and should not provide a unifying perspective encompassing both public discourse and critical or disciplinary discourse.

In rejecting the idea that rhetoric can provide a model for disciplinary discourse, or vice-versa, McGee and Lyne preserve the power of rhetoric as the particular art of seeking adherence in contingent situations, but they also preserve its vulnerability to scorn as an opportunistic substitute for knowledge. In doing so, they accede to the demands of modernism for a scientific method that arrogates reason to itself. In warning that rhetoric will "dangerously oversimplify" the "technological complexities" of "postindustrial society," they seal the separation between the technical and the social with which they began [12, p. 395]. Michael Leff, on the other hand, while acknowledging that the charges against rhetoric are often justified, finds that rhetoric can provide a unifying vision of discourse, that public debate can model scholarly conversation. This model is a "middle ground" between scientism and a rhetoric of perpetual deceit. He calls it "sophistic," for its reliance on concrete practices over abstract principles. Rhetoric becomes a dimension of all discourse, not a type of discourse, and it exhibits the variety of the discourse discussed in this volume. Public discourse and scientific discourse "are characteristically different, not because one employs rhetoric and the other does not, but because they use rather different rhetorics" [13]. Sophistic begins to sound like an ancient variety of philosophical pragmatism, and "the rhetoric of public discourse" no superfluous expression. Both of these essays are closely argued and well informed, about both contemporary and historical matters, and for the rhetorician are the most rewarding in the book.

The sophistic project is taken up in Charles Bazerman's *Shaping Written Knowledge* [14]. The book begins with a practical problem, a pedagogical one, proceeds by a series of detailed examinations of specific practices, and retains as a general premise that the whole point to studying how discourse works is the practical aim of engaging in discourse and doing it better. It provides what is perhaps the first sustained, coherent, historical study of scientific discourse. The pedagogical impetus for the project was the requirement to teach college students to write for their courses in a variety of disciplines; this responsibility led to an investigation of writing within disciplines, to a survey of the resources available to conduct such investigation, to a realization of the complexity of the enterprise, and finally to a focus on the historical development of the experimental report. The twelve chapters are organized into five parts: introductory issues, the early emergence of the experimental article, recent developments in experimental physics, the diffusion of the experimental report model to the social sciences, and concluding chapters on the relationships between science, language, and the practice and teaching of writing.

As Bazerman says in the introduction, his "concern for actual practice leads to a smaller role for rhetorical theorists than is usual in rhetorical histories. The actual writers of scientific texts take center stage" [14, p. 15]. This seems only fair, since the theorists have ignored the discourse in which he is interested. Although the writers are on center stage, in the wings are the resources of a variety of disciplines that provide conceptual assistance—philosophy of science, sociology of science, and language and linguistic theory most prominently. Bazerman's methods vary from chapter to chapter: he draws on the literary tradition of close textual reading, on ethnographic interviews, on statistical analysis of quantified linguistic features, on detailed historical research. In spite of this diversity, Bazerman ties his effort, at the end, to rhetoric and gives a challenge for further serious study:

> Rhetoric has only recently begun to take up the challenge of scientific use of language . . . . We need thoroughgoing and wide-ranging research into the historical and current rhetoric within the sciences and other knowledge-generating communities to gain a grasp of the range of practices, the thematic interactional concerns, the local emergence of typified forms and actions, and the implications for socially produced knowledge [14, p. 32].

Bazerman's own work suggests what the rewards of such study might be.

His ultimate address to rhetoric, rather than to any of the other disciplines he relies on in the process of his analysis, is justified by a conclusion that each of his separate studies leads to: "persuasion is at the heart of science, not . . . the unrespectable fringe." And this persuasion is of the most noble kind to which rhetoric can aspire—it is "a serious method of truth seeking" [14, p. 321]. Bazerman's historical work explains the fulfillment of Henry Oldenburg's realization that "science needed to be agonistically structured" [14, p. 130]. As the secretary of the newly formed Royal Society of London, Oldenburg, not himself a scientist, took it upon himself to develop and promote correspondence with scientists, which he published in the Transactions. Publication led to increased sharing of information, synthesis of work in similar areas, and, inevitably, conflicts between observations and interpretations. By examining 1000 articles in the Transactions, selected from the years 1665 to 1800, Bazerman shows the experimental report to be an historical creation, developing from its epistolary origins through four stages into the report pretty much as we know it. What change over time are the nature and function of "experiment," the amount of attention given to experimental methods, the completeness of results reported, the relative focus on newsworthy facts or on contested issues. Another chapter shows the influence of Sir Isaac Newton, whose intolerance for disagreement led him to develop what Bazerman characterizes as a "logical and empirical juggernaut" in Book 1 of his *Opticks* [14, p. 121]. The force of this method of argument, which "reduces disagreement to error," "provided a model for the form of scientific argument that influenced all of scientific practice" [14, pp. 118, 317]

Bazerman's historical work shows clearly that the scientific report is not a transparent or obvious representation of anything but a carefully adapted method of disciplined argument within a community. He is interested in how both the social structure of the community and the cognitive content of a science are related to its discourse. A community, he believes, "constitutes itself in developing its modes of regular discourse" [14, p. 79]. The development of science as an enterprise committed to "empiricism" presupposed *and* provoked the development of the experimental report as a discourse genre. The use of highly codified language that incorporates the cognitive content and structure of the discipline, the standard topics of organization, the embedding of prior discourse by reference, the stance of the author—all these argumentative resources are marshalled to create empirical accountability to colleagues in a cooperative but agonistic relationship. And by studying not only scientific texts but their social contexts, Bazerman is able to show "how complex a social activity empiricism requires for its realization" [14, p. 149]. This complexity, which he details in sociological terms, suggests why he found the "technology" of classical rhetoric inadequate (or at least inappropriate) for the "knowledge-generating disciplines": classical rhetoric was developed to account for discourse in a different, and perhaps simpler, social context [14, p. 332].

Although he puts to work the conceptual resources of a variety of disciplines, Bazerman interrogates each of them as he goes, intimating throughout what an adequate rhetorical technology for science would have to be like. First, he relies heavily on the extensive sociological studies of science, finding, as I suggested above, that sociology helps to characterize the complexity of the communication system. But he also notes that the sociological work is ahistorical; it tends to emphasize the conflicts and competition among scientists, ignoring the constraints posed by language, convention, and conceptual content, all of which embed the historical development of the discipline. Second, he uses techniques and assumptions from literary study, especially the methods of close reading of written texts. Chapter two, for example, examines texts from three disciplines to discover how each text brings together and contains four contexts: the object of inquiry, prior literature, audience, and the author's self. Yet he must also insist, against the literary tradition, that "nonfiction . . . presents serious literary questions of the representation of worlds in words" [14, p. 60].

Third and most fully, Bazerman adopts concepts from linguistics and sign theory, yet finds neither able to explain the connection between texts and the world. The possibility of reference, in fact, he sees as central to scientific discourse: "the scientific enterprise is built on accounts of nature" [14, p. 60]. His major concluding chapter attempts to remedy some of the inadequacies of linguistics and sign theory with a description of "how language accomplishes the work of science" [14, p. 291]. Traditional linguistics, he argues, has assumed a perfunctory and simplistic correspondence theory of reference, and postmodern sign theory has cut off the linguistic code from reference to a world outside itself. Building from the work of Lev Vvgotsky, Bazerman develops a social-constructivist approach to reference, which turns reference into accountability. In his words:

Science . . . has taken empirical experience as its major touchstone, so that in the process of negotiation of meaning, empirical experience not only constrains the range of possible meanings but is actively sought in the attempt to establish stable meanings from the negotiation. Thus, whatever may be the source of statements, the fate of statements depends on the experience generated by them. In this way science has made nature its ally [14, p. 312].

Earlier, Bazerman admits that he had hoped to avoid the whole question of referentiality, but he finds the issue central to the historical and critical investigations that form the substance of the other chapters [14, p. 187]. His project requires that the social dynamics of science and its allegiance to empirical constraints be productively connected and comprehended by language. It may be, however, that the key notion of "accountability," which describes both the obligations of a scientist to other scientists and the obligations of a scientist to his or her own observations, simply begs the question of reference. It permits Bazerman to continue using the language of correspondence: "precision, accuracy, and clarity," "tightness of fit" [14, pp. 223, 233]. But it also permits him to ascribe these qualities to the perceptions of relevant audiences, to, in other words, the persuasive value of the language chosen by the scientist: "tightness of fit" is a matter of convention. If this is the case, can he say any more than what postmodern sign theory holds, which is that reference is a fiction? He seems to want the matter both ways: "Although the reference is not absolute in the sense of a one-to-one correspondence with objects having self-evidently natural and unchangeable designations and divisions, the reference is more than a literary fiction" [14, p. 224]. What makes the difference here, and what perhaps needs to be more fully worked out, is the emphasis on concrete, practical experience, which, he holds, "keeps reference alive and makes language capable of interacting with the physical world" [14, p. 225]. He shows in the chapter on Arthur Holly Compton that "correspondence" between language and empirical work is not a fiction to Compton but a deliberate, and difficult, construction, one that is always judged by reference both to Compton's own knowledge of what he has done and seen and to his knowledge of other scientists' expectations and the issues and theories that have already been coded by the community. This emphasis on the epistemic power of the concrete also seems appropriate for a project that I characterized above as "sophistic."

If Bazerman, then, has outlined a new approach to the problem of reference, it is one with a rhetorical flavor, given by its reliance on community, convention, and persuasion. He has also sketched out the relationship between a new rhetoric and the contemporary and postmodern disciplines—linguistics, sociology, literature. And he has challenged rhetoric to take responsibility for the history and criticism of an arena of discourse that it has largely ignored. In fact, Bazerman's work can be read as the beginnings of an essentialist argument against the traditional separation of rhetoric and science: science must be rhetorical, in the most basic senses of rhetoric. I say the beginnings because one basic sense is omitted, and that is the public dimension of rhetoric. In Bazerman's book the scientific community stands for the public realm; the knowledge-generating disciplines are not also seen as power-wielding entities in the larger arena of public discourse. The relationship between power and knowledge is a recurrent issue in postmodernism, one that a full rhetoric of science must comprehend. As McGee and Lyne reminded us, the danger to rhetoric in attending to knowledge is that power may be relinquished.

The Wisconsin series is a timely project. These three books illustrate the pitfalls and prospects for rhetoric that the project represents. The atrophy of modern rhetoric may, indeed, be at an end, and it is the postmodern shift that makes projects like this promise not malignant hypertrophy but healthy revitalization. McKeon's approach to the history of rhetoric in the middle ages seems also to describe what we can find in this moment: a history that "might give significance and lively interest to the altering definitions, the differentiation of various conceptions of rhetoric itself, and the spread of the devices of rhetoric itself to subject matters far from those ordinarily ascribed to it" [2, p. 124]. In this way, we see rhetoric as McKeon saw it, "a continuing art undergoing revolutionary change.

# REFERENCES

1. B. Vickers, The Atrophy of Modern Rhetoric: Vico to de Man, *Rhetorica, 6*, 1988.
2. R. McKeon, *Rhetoric: Essays in Invention and Discovery*, M. Backman (ed.), Ox Bow Press, Woodbridge, Connecticut, 1987.
3. M. C. Leff, In Search of Ariadne's Thread: A Review of the Recent Literature on Rhetorical Theory, *Central States Speech Journal, 29*, 1978.
4. D. N. McCloskey, *The Rhetoric of Economics*, University of Wisconsin Press, Madison, Wisconsin, 1985.
5. A. Klamer, As If Economists and Their Subjects Were Rational, J. S. Nelson, in *The Rhetoric of the Human Sciences: Language and Argument in Scholarship and Public Affairs*, J. S. Nelson, A. Megill, and D. N. McCloskey (eds.), University of Wisconsin Press, Madison, Wisconsin, 1987.
6. J. S. Nelson, A. Megill, and D. N. McCloskey (eds.), *The Rhetoric of the Human Sciences: Language and Argument in Scholarship and Public Affairs*, University of Wisconsin Press, Madison, Wisconsin, 1987.
7. R. Rorty, Science as Solidarity, in *The Rhetoric of the Human Sciences: Language and Argument in Scholarship and Public Affairs*, J. S. Nelson, A Megill, and D. N. McCloskey (eds.), University of Wisconsin Press, Madison, Wisconsin, 1987.
8. P. J. Davis and R. Hersh, Rhetoric and Mathematics, in *The Rhetoric of the Human Sciences: Language and Argument in Scholarship and Public Affairs*, J. S. Nelson, A. Megill, and D. N. McCloskey (eds.), University of Wisconsin Press, Madison, Wisconsin, 1987.
9. J. Campbell, Charles Darwin: Rhetorician of Science, in *The Rhetoric of the Human Sciences: Language and Argument in Scholarship and Public Affairs*, J. S. Nelson, A. Megill, and D. N. McCloskey (eds.), University of Wisconsin Press, Madison, Wisconsin, 1987.
10. J. S. Nelson, Stories of Science and Politics, in *The Rhetoric of the Human Sciences: Language and Argument in Scholarship and Public Affairs*, J. S. Nelson, A. Megill, and D. N. McCloskey (eds.), University of Wisconsin Press, Madison, Wisconsin, 1987.
11. J. Lyne, Rhetorics of Inquiry, *Quarterly Journal of Speech, 71*, 1985.
12. M. C. McGee and J. R. Lyne, What Are Nice Folks Like You Doing in a Place Like This? Some Entailments of Treating Knowledge Claims Rhetorically, in *The Rhetoric of the Human Sciences: Language and Argument in Scholarship and Public Affairs*, J. S. Nelson, A. Megill, and D. N. McCloskey (eds.), University of Wisconsin Press, Madison, Wisconsin, 1987.
13. M. Leff, Modern Sophistic and the Unity of Rhetoric, in *The Rhetoric of the Human Sciences: Language and Argument in Scholarship and Public Affairs*, J. S. Nelson, A. Megill, and D. N. McCloskey (eds.), University of Wisconsin Press, Madison, Wisconsin, 1987.
14. C. Bazerman, *Shaping Written Knowledge: The Genre and Activity of the Experimental Article in Science*, University of Wisconsin Press, Madison, Wisconsin, 1988.

# CHAPTER 4

# Feminist Critiques of Science and Gender Issues

Feminist critiques of technical communication and studies of gender issues in technical communication have been relatively few, though this state of affairs is rapidly changing. Perhaps the most widely known are those of M. M. Lay and J. Allen [1, 2]. Due to this limited amount of material specific to technical communication, much of my discussion will concern feminist critiques of science, the translation to technology and to technical communication being fairly straightforward.

After defining and outlining the theory of feminist critiques of science, I will review the literature on this general topic. I will then introduce the reprinted articles concerning technical communication specifically.

## DEFINITION AND THEORY

As with the other topics of this book, the discussion of feminist critiques of science faces problems of definition. The careful exploration of the difficulty in being definitive is revealing, however, because that is partly the point of those feminist critics who reject the scientific insistence that valid knowledge can only be absolute, single, and incontrovertible.

### Indefiniteness

One of the difficulties in discussing feminist critiques of science definitively is that humanistic topics such as this are inherently open-ended. Indeed, it is the very nature of humanism itself never to be finally formed. Another reason is the centrality of language to these topics, language which both reflects and shapes the culture in which it occurs. An additional reason is the multiplicity of feminist voices, some of which differ pointedly from others: there simply is no univocal feminist position on science.[1]

## General Features of Feminist Critiques of Science

Though there is difference and debate about feminist critiques of science, many of its features can usefully be clustered into two broad groupings: the narrow exclusiveness of the scientific mentality and, relatedly, the gendered nature of science.[2]

*Narrowness*

A principal criticism by feminists focuses on the basic frame of mind of science. This scientific mentality, it is argued, strongly prefers observing the external, non-human world rather than the social, psychological, and political world of human behavior and interrelation. This preference can pointedly exclude very human issues such as generating rigorous scientific observations about our society rather than arguing about how best that society should be conducted. Or it could be concerned with improving the accuracy of nuclear warheads rather than grappling with the question of whether it is ever ethical to use such weapons or even build them. N. Chodorow, for instance, explains that the basic frame of mind which views one's self as separate from nature is the product of male psychodynamics in societies in which women are the primary care-givers [5]. This frame of mind emphasizes the external and objective.

This objectivity can be damaging when the scientific gaze is turned to humankind. Such excesses of objectivization, furthermore, are usually most heavily felt by the disadvantaged and disempowered. The infamous Tuskegee experiments in the 1920s in which African Americans diagnosed as syphilitic were treated only with a placebo so that the unimpeded course of the disease could be studied is a good example of a "scientific experiment" that would never have been carried out on whites. Similar abuses occur for all relatively powerless groups, feminists contend. Related to such excesses is the valorization of objectivity itself, critics such as Lay hold, an elevation which seems almost necessarily to devalue subjectivity, and with it feelings, principles, intuitions, even one's sense of self, all of which cannot be objectified and so suffer either neglect if or denial by scientists [1, p. 355].

*Gendered Nature of Science*

Another principal feminist criticism of science concerns the gendered nature of science, whether in principle or in practice, specifically its maleness which has

---

[1] Though Lay indicates six common characteristics of feminist theory, she adds, "Feminist theorists . . . resist a uniform definition of feminist studies to avoid stereotyping women . . . [1, p. 350].

[2] See Lay for an alternate grouping of characterists of feminist theory [1].

operated to disadvantage and exclude women. This maleness stems from the origins of science in male-dominated ancient Greek culture and in later male-dominated seventeenth- and eighteenth-century Euro-American culture. It is revealed in the valorization of such principles as objectivity, impersonalness, logicalness and linear thinking, obsession with hierarchy (as in "the natural order"), and the disdain of feelings.

This maleness is also indicated, critics hold, in the goals of science to predict and control. This prediction and control is not always in the beneficial sense of predicting the tides so as to prevent damage and injury, but also in the pejorative sense of domination. Such domination seeks the control of nature, which is often personified in female form as the primal mother. Thus male Science is taken as dominating and controlling not only female Mother Nature but by extension all of womankind as well as the "female" attributes of sensitivity and caring.[3]

Some of these general representations, to be sure, are questioned by other women. H. E. Longino and E. Hammonds, for example, noted that some women scientists respond to feminist critiques of science by asserting the pragmatic effectiveness of science and affirming its capacity to neutralize biases [6]. Longino and Hammonds believe that these women scientists have become too personally invested in the pursuit of conventional masculinist science to recognize their distance from women's interests.

Nevertheless, the historical *de facto* near-total exclusion of women from the pantheon of Science remains and calls out to be examined, explained, and rectified. This rectification will require the conscientious reconsideration of what exactly science is, of how the principles of science differ from the practice, of how science historically both shapes and is shaped, and the social importance to be accorded scientific knowledge.

## LITERATURE REVIEW OF FEMINIST CRITIQUES OF SCIENCE

A good review article of developments in feminist critiques of science from 1978 through 1988 is *A Decade of Feminist Critiques in the Natural Sciences* by R. Bleier, the renowned neurophysiologist and prominent feminist [9]. Bleier insightfully explains why, despite major feminist impacts in so many other fields, feminism has not as yet had substantial impact on science and so few women have become important scientists. It is because, Bleier says, science, if not innately at

---

[3] Some feminist critiques of science go beyond science itself. Feminist criticism informed by recent French philosophers (e.g., Foucault, Derrida, Deleuze, Lyotard) sees science as reflecting, maintaining, and reproducing a prevailing social-economic-cultural power structure that is androcentric and inimical to women. Such criticism is principally concerned with undermining this social order and only derivately with critiquing science specifically.

least historically, has been a highly gendered activity that valorizes positivism, objectivity, and value neutrality, all principles which feminism calls into question. More particularly, science as a cultural practice has distanced itself from political activity and political responsibility so much that political activism by scientists is seen as anti-scientific.

A good bibliography in this area can be found in A. Wylie, K. Okruhlik, L. Thielen-Wilson, and S. Morton's *Feminist Critiques of Science*, which reviews the epistemological and methodological literature through 1988 [10].

Three major collections of feminist critiques of science have been published in recent years, edited respectively by R. Bleier, by N. Tuana, and by M. Hirsch and E. Fox Keller. Among these three collections, there is a consensus that many abuses in the name of science have occurred and have been suffered by women heavily (though not exclusively). These abuses include excesses of impersonalness by which people are treated insensitively as blank material objects, even dehumanized; the use of science improperly to predict and control as a sort of domination; the implied invalidation of non-scientific knowledge and attitudes; the selection of topics of scientific research and the interpretation of research findings in ways that exclude, disadvantage, disempower, or devalue women; the valorizing of scientific, specialized knowledge so as to exclude non-specialists from a role in making important decisions; and, perhaps most fundamentally, the predominance of men in science, suggesting that science (in practice if not in principle) is a characteristically male activity that, therefore, excludes or disadvantages women while automatically advantaging men.

Though there is consensus on the above matters, there is nonetheless a good deal of difference of opinion among these collections on other basic matters such as the definition of science itself, whether it is characteristically or only circumstantially masculine. Other opinions differ as to the ultimate worth of science vis-a-vis feminism, wondering if a good feminist can also be a good scientist. Still other opinions differ as to the basic worth of objectivity in light of its implicit rejection of subjectivity or the implicit subjection in treating people as subjects.

Despite these differences, however, the feminist critique of science, while pursuing its own purposes, clearly resonates with parallel developments in other areas. Such parallels include relativity and indeterminacy in physics, studies of the sociological basis of scientific knowledge, ethnomethodology in anthropology, and culture criticism in literary studies. Indeed, such developments are so pervasive as to be characterized collectively as a movement, namely postmodernism. For our purposes, the general thrust of these developments is to question, if not challenge, the traditional privilege accorded to science and its practices, methodologies, and underlying assumptions, and by extension, technology.

## R. Bleier's (ed.) *Feminist Approaches to Science*

Bleier's collection covers a range of views on feminism and science [11]. In her own essay Bleier generally opposes science and disbelieves in the possibility of a science in which feminists can participate in good conscience. She holds that the practice of science often (if not necessarily) results in exploitation and destruction of the objects of scientific studies, whether they be Nature, animals, or people.

M. Namenwirth in Bleier points out that many of the concepual paradigms and research interests of science as commonly practiced serve the interests of the white male middle class. She indicates, however, that good science can be done by both men and women, though with the proviso that other sorts of knowledge more personalistic and emotionally sensitive must also be granted authority. This is because men and women share the same abilities and capacities; differences between them are not of kind but only of degree.

E. Fee in Bleier relates feminist critiques of science to (Marxist) critiques of science based on class, race, and all other expressions of power. She explains that a definitive feminist critique of science is not possible because of the variety of feminisms. In addition, she calls for feminists to become active in science principally to advance feminism, the result being, one imagines, an interdigitation of feminist science and masculinist science that would expand the definition and range of science as a whole. Thus though there is not a specifically feminine science, science can be practiced in the service of feminist interests.

H. Rose in Bleier opposes traditional science on many grounds including the rape and destruction of nature; the objectivizing of women's activities so as to trivialize and dismiss them; the atomization of our holistic interdependency with nature; and the narrowness of linear thinking. Rose devotes particular emphasis to caring, pointing out that caring cannot be reduced to words, mechanical processes, or abstractions. The apparent implicit essentialism in this position would seem to clash with the strong social constructionism of other feminist critiques. The strong social constructionist position holds that behaviors such as caring are amenable to change and therefore are not innately specific to either men or women.[4]

---

[4] The role of social constructionism in feminism is an important crux. M. Hirsch and E. Fox Keller in *Conflicts in Feminism* explore some of the difficulties in taking the radical social constructionist position. They also explore the difficulties of denying biological sex as a determinate of men's or women's behaviors. Nonetheless, they acknowledge the cultural and political usefulness of holding a social constructionist position in keeping one from being bound by the past [13]. J. Lorber and S. A. Farrell in *The Social Construction of Gender* also explore the significance of social constructionism for feminism [14].

## N. Tuana's (ed.)
### Feminism in Science [16]

Tuana in her introduction lists the many positive contributions of feminist critiques of science in recent years including equity studies showing discrimination in education and employment. Feminist scientists have also helped to identify the psychological and social mechanisms of discriminations. Still others have probed science itself, questioning the historical value-ladenness and bias of science or introducing new epistemologies providing alternative understandings of social experience [17].

Interestingly, Tuana also briefly mentions difficulties with a radical social constructionist position, citing various feminists taking a radical constructionist view, others taking a biological deterministic view, and still others taking a mixed, interactionist view. The social constructionist position of some feminists on gender differences implies that those differences might either be obliterated or selectively emphasized depending on social policies. In contrast, Tuana continues, other feminists such as Elshtain and MacMillan reject the social construction of gender in favor of the biological construction of gender. Biological determinism as espoused by these feminists proposes that women have gender-based traits that are to be valued and that grant a superiority to women. But, says Tuana, Birke sees biological determinism as fraught with the same problems whether proposed by feminists or anti-feminists. Tuana herself takes a middle-of-the-road feminist position acknowledging that biology plays a significant role in behavior but only in interaction with equally-significant social factors.

S. Harding in Tuana addresses the question, Is there a feminist science? Her answer is, No. Among her reasons are the implied repudiation of important feminist knowledge gained through traditional scientific methods (e.g., C. Gilligan on differences in morality and N. Chodorow in psychology). Harding thus does not categorically reject science; indeed, science for Harding would seem to need feminist scientists for a more balanced, fuller picture of our selves and our world. She notes, for instance, that some of the findings of feminist scientists are more objective specifically on the basis of the feminist politics driving these studies.[5] Thus for Harding, objectivity and the pursuit of truth (or at least the minimization of falseness) are important values which cannot of themselves be the basis for feminists repudiating science. Though she rejects the possibility of a distinctly

---

[5] Lay cites Harding's observation that informing the practice of science with a feminist "subjectivity" will, paradoxically, make science more objective [1, p. 351].

feminist science, Harding emphasizes that feminists do bring important interests, insights, and information to science.[6,7]

H. E. Longino in Tuana considers the question, Can there be a feminist science? [20]. She decides that the question misses the point and instead one should ask, How can one do science as a feminist? One reason for Longino's reconceptualization is that to reject one approach as incorrect and fallaciously absolutist and then to embrace another approach as absolutely true is simply to re-commit the fallacy of absolutism. Bias in science is not necessarily damaging, Longino surprisingly notes. She argues that the practice of science is embedded in the prevailing culture, indeed, is inescapably shaped, biased, and conditioned by it. The practice of science cannot but be biased; the task is to compensate for biases in a socially responsive way. Thus Longino, though calling for science to be accountable according to certain social goals, does not categorically reject science but argues for a new form of science consistent with feminist principles and goals.

E. Fox Keller in Tuana presents an important discussion valuable for its breadth, complexity, and synthesizing power [21]. She develops a sophisticated exploration of a fundamental yet possibly unanswerable question: Is sex to gender as nature is to science? The primary thrust of Keller's essay is to establish the importance of the middle ground between science and nature, gender and sex so as to avoid the pitfalls of the polar, dualistic extremes of absolutism and relativism on the one hand, and essentialism and radical social constructionism on the other.[8] Defining this middle ground is difficult, though, not only because of the inherent conceptual difficulty but also because of the strong social pressure in favor of one pole or the other.

On the possibility of a distinctly feminist science, Keller is guarded, partly because the very conception seems to reinforce damaging, traditional stereotypes and because it seems to diminish the work of women scientists already done. This in turn fosters the continuing exclusion of women from science. For Keller, it is not a question of either sex *or* gender but both sex *and* gender; the same for empirical science and social constructionism.

---

[6] E. A. Buker notes that Harding differs from some postmodern feminists who hold politics as the foundation for theory [19]. Buker says that Harding, in affirming empiricism and objectivity, avoids many of the problems of relativism [19].

[7] S. Harding has written several books on feminism and science. *Whose Science? Whose Knowledge?* (1991) moves beyond her earlier landmark book, *The Science Question in Feminism* (1986) [29, 30]. The earlier book raised feminist objections to traditional science and so was generally reactive, while the later book looks toward constructing a sort of science specifically congenial to feminist interests.

[8] Lay, for instance, grapples with the question, "Where should the origin of difference be located, within biology or society?" [1, p. 355].

L. Heldke in Tuana develops a synthesis of the views of E. Fox Keller and the pragmatist philosopher J. Dewey [23]. This synthesis yields a holistic view of science which avoids many of the features of traditional science criticized by feminists and which recognizes the power of language to constitute both knowledge and communties. Heldke sees Dewey and Keller as sharing a common epistemological tradition, a common ontology, and a common principle of inquiry. She develops what she calls the "co-responsible option" regarding the interrelation of science and nature. This position rejects both the absolutism of, say, positivism in its claim for an independent, antecedent world and the relativist alternative holding that there is no ground for knowledge outside ourselves. The epistemologies of Dewey and Keller, Heldke says, free us from the dualism implicit in traditional epistemologies.

For Dewey, inquiry is constituted by a relationship between the inquirer and the inquired. Thus knowing is always provisional and contingent, never absolute. Keller's view, Heldke points out, like Dewey's, involves an interactive communication. For Keller, both the world and the knower interact so that our knowledge is neither entirely discovered nor entirely constructed. For both, the external, material world does not of itself constitute knowledge as much as it conditions knowledge by defining the limits that cannot be ignored to our socially constructed knowledge.[9]

In reviewing these two collections, Bleier's and Tuana's, one can easily become confused if one expects crisp, clear answers. There are various definitions of science; there are various understandings of the relation of feminists to science; and there are various attitudes toward social constructionism and objectivity. This indefiniteness and multiplicity of positions is *not*, however, a deficiency. Indeed, it is, many feminists hold, precisely the nature of a feminist perspective not to be confined inflexibly to distinct, absolutist definitions and not to be constrained by any single dogmatic, authoritative voice. Freedom and open-ended, progressive evolution is, in large measure, the hallmark of feminists' conceptualization of womankind and the world.[10] Conflict in feminism, in fact, is the topic of the next collection of essays.

---

[9] R. Rorty's *Inquiry as Recontextualization: An Anti-Dualist Account of Interpretation* outlines Dewey's holism while urging the abandonment of traditional dualistic thinking [24]. Rorty's, Dewey's, and Keller's middle ground positions are difficult to articulate, however, precisely because they oppose the easy categorizations of dualistic thinking.

[10] S. Gorelick, for instance, in *Contradictions of Feminist Methodology* critiques some feminist writing for its elitism and false consciousness yet returns to affirm a basically feminist stance in culture criticism albeit on an improved level [25].

## M. Hirsch and E. Fox Keller (eds.)
### Conflicts in Feminism

This collection of essays explores the vital importance of conflict both internally within feminism and externally with other ideologies such as science [26]. The debate about feminism even within feminism is not so much a shortcoming as a confirmation of the open-ended potential of women defining womanhood. Though the essays deal for the most part with feminism in general, several essays deal with science and technology directly.

J. Acker in Hirsch and Keller explores the gendered nature of organizations, identifying and critiquing the importance of the principle of abstraction to traditional complex organizations [27]. Many of her criticisms hold as well for abstraction as a principle of science. Both science and complex organizations are founded on the principle of abstraction by which the uniquely personal is specifically displaced by impersonalness. The disengagement of the researcher from his or her own feelings likewise is highly valued. In scientific research, for example, the particular person performing the research should be irrelevant to the results of the research and the research should be replicable by any other careful researcher.

F. M. Cancian exemplifies the difficulty in articulating a manner of doing science that avoids dichotomizing science and feminism [28]. She advocates a feminist methodology. It consists of five points that emphasize the social context in which scientific research is practiced and the social interests it serves. Her social practice method of science continues to use conventional, quantitative research methodologies on a limited basis while flexibly intermixing qualitative methodologies to emphasize the social aspects of research. Thus feminist scientists need not run the risk of radically separating themselves from traditional bases for authority. Cancian's position is not to reject objectivity but to advocate additional scientific standards and methodologies that equality. In addition, Cancian says, the power structure of academia should be challenged in its systematic valorization of knowledge of only a scientific sort.

Wylie et al. attempt to summarize the current state of research on feminism and science, indicating that there does not appear to be a set of cognitive capacities unique to women that would warrant a distinctly feminist science. Rather than reacting against patriarchal biases in science, feminists should redirect the conduct of science toward feminist social and political interests.

The net insight gained from these three books concerns the sexism of science as a social and intellectual practice. These criticisms reinforce the thrust of the other topics of this book, they question the very meaning of "science" and "technology," and they stimulate us to accommodate historically-excluded people, their knowledge, and their lives.

## LITERATURE REVIEW OF GENDER ISSUES

Recently (December 1992), *IEEE Transactions on Professional Communication* published a special issue on gender. Though many of the articles have a fairly narrow focus, the introductory essay by B. A. Sauer presents a broad overview of the topic [31].

Sauer underscores the pragmatic importance of gender issues in the professional world. The ramifications of the increased participation of women in the professional, scientific, and technical world are manifold, from requiring new policies on child care to adjustments in compensation programs. But these overt developments can distract us, she adds, from equally important covert issues, in particular the silent powerful role of language itself in shaping not only our communications but our selves and our society. Sauer also reiterates that the gendered nature of language has powerful, pervasive ramifications. The goals of feminist theorists of technical communication, therefore, are to reveal the hidden assumptions, power structures, and ways of thinking underlying communications. The identification and rectification of these matters would, Sauer explains, yield a professional workplace more effective, efficient, and gender-neutral. They could also yield a fundamental redefinition of the very problems and solutions we communicate about.

S. Dragga's *Women and the Profession of Technical Writing: Social and Economic Influences and Implications* examines the causes of the dominance of women in the profession of technical writing [32]. This dominance holds the threat of depreciating the field, Dragga explains, using the same principles and historical observations cited by J. Allen (reprinted here). Dragga clearly describes the damaging circular reasoning historically associated with women coming into prominence in any field (also described by Allen) by which it become less valued, less well-paid, and less attractive.

Dragga urges that we as a profession take steps to prevent this prejudiced depreciation. He recommends the publication of results from periodic detailed surveys of wages and working conditions by the Society for Technical Communication and the Association of Teachers of Technical Writing. This will allow technical writers to objectively assess their situation relative to the national picture. In addition, special committees should establish national guidelines for wages and working conditions to allow technical writers to identify when they are being victimized. These committees would also investigate claims of discrimination and publish lists of offenders. Dragga also urges that technical writing teachers discuss in their classes the potential for depreciation and present specific expections of wages and working condition to their students.

M. M. Lay's *Gender Studies: Implications for the Professional Communication Classroom* is a succinct statement of the characteristics of feminist critiques of science relating to professional, technical, and scientific communication [33]. Lay calls for a change in the professional communication classroom from its traditional character. She sees the professional communication classroom (which encompasses technical and scientific communication) as strongly shaped by traditional gender roles and heavily biased in favor of men and against women. Logic, rationality, objectivity, even science itself are understood by Lay as characteristically male ways of thinking and viewing the world while competition, autonomy, and individuality are characteristically male ways of interacting socially. These male characteristics so intimately inform the traditional classroom of higher education that they practically define the norm. As a result, men excel in these classrooms and carry these values and behaviors into the non-academic professional world.

Women, Lay explains, are more concerned with feelings, intuitions, and non-hierarchical ways of understanding the world and prefer to interact socially in ways that preserve relationships, strength groups bonds, and value continued communication for its own sake. As a result, in the traditional classroom, women are disfavored, excluded, and forced either to compromise or to deny their gender identity.

To remedy these historical inequities, Lay proposes a pointedly feminist classroom. This involves revamping learning, teaching, reading and writing, speaking, and group activities along lines favorable to women.

## REPRINTED ARTICLES

### M. M. Lay

Lay's *Feminist Theory and the Redefinition of Technical Communication* seeks to redefine the field of technical communication in light of feminist theory in order to reflect important alterations in attitudes and activities [1]. Due to the scope of her topic, Lay sketches only the broad outline of these changes. She first establishes the six characteristics of feminist theory and the three issues that divide feminists, then redefines technical communication on the basis of these factors, yielding three principal effects.

The six characteristics Lay discusses, gleaned from a wide array of feminist scholarship, are the celebration of difference; a coherent theory activating social changes; the acknowledgment of the scholars' background and values (in shaping the scholars' perceptions); the inclusion of women's experiences; the study of gaps and silences in traditional scholarship; and new sources of knowledge which refuse to be confined to a particular canon.

The three issues dividing feminists are the emphasis on either similarities or differences among women and men; the basis of these differences either in culture (social constructionism) or in biological traits (essentialism or genetic determinism); and either the promotion or the discouragement of binary opposition between women and men. These issues will likely never be resolved, Lay points out, and so will continually complicate discussions of feminism and technical communication and will preclude definitive statements on this topic.

Lay's scholarship redefines the field in ways that reflect feminist insights and affirm women in general. The first effect is to refute the myth of objectivity, which implicitly devalues all aspects of subjectivity. The dehumanizing that can go hand in hand with treating people as the *objects* of study holds great potential for abuse. The second effect is to expand what is meant by "science" to include recognition of the social nature of scientific knowledge and communication. The third effect is the movement toward increased collaborative writing and social constructionism (see also Lay's collaborative writing article below [34]). Historically, the technical writer has been understood as an individual writing, as an isolate. The social constructionist position, on the other hand, has the communicator playing a role in the very constitution of knowledge, a role both more active and more powerful. At the same time, it empowers and validates voices, particularly women's, that have historically been excluded from the shaping of valued knowledge.

In an earlier article, *Interpersonal Conflict in Collaborative Writing*, Lay explores some of these practical difficulties of traditional male-centered technical communication and presents collaborative writing as the feminist solution to them [34]. For Lay, collaborative writing is the paradigm activity of the technical communication of the future. Technical writing, Lay says, should be done collaboratively simply because technical knowledge itself is socially, collaboratively constructed. In addition, the rights of the historically disempowered or excluded insist that they now be integrated in a powerful, meaningful way in all communications. This accommodation will, in turn, involve entirely new, more egalitarian and socially inclusive ways of interacting.

Lay raises in this article fundamental questions facing the call for collaborative writing on the basis of gender-fairness or gender-neutrality. Should women write and act more like men? Should men write and act more like women? Or should both women and men write and act toward some neutral ground of androgyny, which is Lay's suggestion. Underlying these questions is the more fundamental question of the wellspring of human behavior: Is it culture (social constructionism), or genetics (biological determinism), or an interactive, alterable admixture of the two? These questions, as we have seen, cannot easily be answered, though the need for more equitable communications is clear.

## J. Allen

Allen's *Gender Issues in Technical Communication Studies: An Overview of the Implications for the Profession, Research, and Pedagogy* has a focus different from Lay's [2]. Rather than focusing on the nature of technical communication in theory as Lay does, Allen focuses on the actual practice of technical communication in its gender-unfairness and inequities. Allen discusses these shortcoming from the perspective of basic principles of modern business and management. By discriminating against, excluding, and devaluing the work of women, the practice of technical communication fails to take advantage of an important resource and so hampers its own creativity, efficiency, and productivity.

Allen's review is particularly important for two reasons. First, it reviews a broad range of careful investigations of the field itself. Thus it informs the reader about the work of others as well as Allen while it bases its claims on actuality rather than theorization or speculation. Second, it reveals the great complexity of gender studies. One cannot, for example, simply use salaries as an indication of the worth of the work of communicators because of the interplay of other factors. In particular, many studies in other fields have shown that as more and more women enter a field, the general impression of the value of the work of that field declines even though no substantive change occurs in the work itself. That is, the work is devalued entirely on the basis of its being done more and more by women rather than by men. In addition, Allen is carefully even-handed in presenting counterposing studies and perspectives, and even goes one step further in offering her own insightful criticisms of such controversies.

Allen's findings will sound familiar to anyone who has read much in feminist culture criticism because they examine a pandemic phenomenon. Though there has been a growing "feminization" of the field as indicated by the growing proportion of women technical communicators, these women communicators continue to be paid a lower salary than men counterparts, even when the data are corrected for status and seniority. There has also been an inclination to perceive technical communication as having lower innate value as, and more precisely *because*, the proportion of women in the field increases. In addition, women are not promoted as quickly as men, decisions on promotion often involving biasing factors other than productivity and quality.

Regarding research on gender issues in technical communication, Allen calls for studies re-investigating earlier studies on sexist biases to clarify assumptions and issues and to improve the rigor of the investigations. She also calls for study of the nature of gender differences themselves rather than assuming the preferability of gender neutrality or indifference (in this, Allen differs from Lay). Thus Allen even-handedly seeks to avoid perpetuating stereotypes while also seeking to avoid unwarranted, untested assumptions.

After identifying many aspects of gender-unfairness in the workplace, Allen calls for corresponding, corrective adjustments to technical communication pedagogy so as to prevent the continuance of this unfairness.

Clearly a humanistic approach not only to technical communication but to science and technology themselves would be highly consonant with the criticisms of Lay and Allen. Recognizing the inescapable rhetoricity and social constructedness of science and technology naturally opens the door to more socially responsive shaping of all these enterprises. A humanistic approach would at the same time recognize the validity and importance of feminist critiques and gender issues, and ethically compel the resolution of these matters.

## REFERENCES

1. M. M. Lay, Feminist Theory and the Redefinition of Technical Communication, *Journal of Business and Technical Communication, 5*:4, pp. 348-370, 1991.
2. J. Allen, Gender Issues in Technical Communication Studies, *Journal of Business and Technical Communication, 5*:4, pp. 371-392, 1991.
3. P. Feyerabend, *Against Method*, Verso Books, New York, 1988.
4. R. Rorty, Solidarity or Objectivity? *Objectivity, Relativism, and Truth*, Cambridge University Press, Cambridge, England, 1991.
5. N. Chodorow, *The Reproduction of Mothering: Psychoanalysis and the Sociology of Gender*, University of California Press, Berkeley, California, 1978.
6. H. E. Longino and E. Hammonds, Conflicts and Tensions in the Feminist Study of Gender and Science, in *Conflicts in Feminism*, M. Hirsch and E. Fox Keller (eds.), Routledge, New York, 1990.
7. Plato, *Plato's Theory of Knowledge: the Theaetetus and the Sophist*, F. M. Conford (trans.), Harcourt, Brace, New York, 1935.
8. Plato, *Phaedrus*, W. Hamilton (trans.), Penguin, New York, 1973.
9. R. Bleier, A Decade of Feminist Critiques in the Natural Sciences, *Signs, 14*:1, pp. 186-195, 1988.
10. A. Wylie, K. Okruhlik, L. Thielen-Wilson, and S. Morton, Feminist Critiques of Science: The Epistemological and Methodological Literature, *Women's Studies International Forum, 12*:3, pp. 379-388, 1989.
11. R. Bleier (ed.), *Feminist Approaches to Science*, Pergamon Press, New York, 1986.
12. R. Bleier, Introduction, in *Feminist Approaches to Science*, R. Bleier (ed.), Pergamon Press, New York, 1986.
13. M. Namenwirth, Science Seen Through a Feminist, in *Feminist Approaches to Science*, R. Bleier (ed.), Pergamon Press, New York, 1986.
14. E. Fee, Critiques of Modern Science: The Relationship of Feminism to Other Radical Epistemologies, in *Feminist Approaches to Science*, R. Bleier (ed.), Pergamon Press, New York, 1986.
15. H. Rose, Beyond Masculinist Realities, in *Feminist Approaches to Science*, R. Bleier (ed.), Pergamon Press, New York, 1986.
16. N. Tuana (ed.), *Feminism & Science*, Indiana University Press, Bloomington, Indiana, 1989.
17. N. Tuana, Introduction, in *Feminism & Science*, N. Tuana (ed.), Indiana University Press, Bloomington, Indiana, 1989.
18. S. Harding, Is There a Feminist Method?, in *Feminism & Science*, N. Tuana (ed.), Indiana University Press, Bloomington, Indiana, 1989.
19. E. A. Buker, Rhetoric in Postmodern Feminism: Put-offs, Put-Ons, and Political Plays, in *The Interpretive Turn*, D. R. Hiley, J. F. Bohman, and R. Shusterman (eds.), Cornell University Press, Ithaca, New York, 1991.

20. H. E. Longino, Can There Be a Feminist Science?, in *Feminism & Science*, N. Tuana (ed.), Indiana University Press, Bloomington, Indiana, 1989.
21. E. Fox Keller, The Gender/Science System: or, Is Sex to Gender as Nature is to Science?, N. Tuana, in *Feminism & Science*, N. Tuana (ed.), Indiana University Press, Bloomington, Indiana, 1989.
22. E. Fox Keller, *A Feeling for the Organism: The Life and Work of Barbara McClintock*, W. H. Freeman, New York, 1983.
23. L. Heldke, John Dewey and Evelyn Fox Keller: A Shared Epistemological Tradition, in *Feminism & Science*, N. Tuana (ed.), Indiana University Press, Bloomington, Indiana, 1989.
24. R. Rorty, Inquiry as Recontextualization: An Anti-Dualist Account of Interpretation, in *The Interpretive Turn*, D. Hiley, J. F. Bohman, and R. Shusterman (eds.), Cornell University Press, Ithaca, New York, 1991.
25. S. Gorelick, Contradictions of Feminist Methodology, *Gender & Society, 5*:4, pp. 459-477, 1991.
26. M. Hirsch and E. F. Keller (eds.), *Conflicts in Feminism*, Routledge, New York, 1990.
27. J. Acker, Hierarchies, Jobs, Bodies: A Theory of Gendered Organizations, in *The Social Construction of Gender*, J. Lorber and S. A. Farrell (ed.), Sage, Newbury Park, California, 1991.
28. F. M. Cancian, Feminist Science: Methodologies that Challenge Inequality, *Gender & Society, 6*:4, pp. 623-642, 1992.
29. S. Harding, *Whose Science? Whose Knowledge?*, Cornell University Press, Ithaca, New York, 1991.
30. S. Harding, *The Science Question in Feminism*, Cornell University Press, Ithaca, New York, 1986.
31. B. A. Sauer, Introduction: Gender and Technical Communication, *IEEE Transactions on Professional Communication, 35*:4, (Special Issue on Gender), pp. 193-195, 1992.
32. S. Dragga, Women in the Profession of Technical Writing: Social and Economic Influences, *Journal of Business and Technical Communication, 7*:3, pp. 312-321, 1993.
33. M. M. Lay, Gender Studies: Implications for the Professional Communication Classroom, in *Professional Communication: The Social Perspective*, N. R. Blyler and C. Thrall (eds.), Sage, Newbury Park, California, 1993.
34. M. M. Lay, Interpersonal Conflict in Collaborative Writing: What We Can Learn from Gender Studies, *Journal of Business and Technical Communication, 3*:2, pp. 5-27, 1989.

# FEMINIST THEORY AND THE REDEFINITION OF TECHNICAL COMMUNICATION

## *Mary M. Lay*

Technical communication scholars take an interdisciplinary approach to their field. In addition to theories and methodologies from linguistics, speech communication, literature, anthropology, science, and rhetoric, feminist theory and subsequent gender studies now also influence technical communication research, as well as other disciplines. Women's experiences have become legitimate subjects for study: Women researchers acknowledge their distinct interests as they generate knowledge, social structures have been scrutinized for sexual bias, and scholars have identified women's ways of knowing, communicating, and leading (see Gilligan; Belenky et al.; Helgesen, all of which attempt to describe the distinct ways in which women make ethical decisions, determine knowledge, and manage others [1-3]).

How then has technical communication—either directly or through its affiliation with these other disciplines—been affected by feminist theory and gender studies? Defined initially as the objective transfer of information, technical communication has long been privileged in its affiliation with science and technology. Now, however, feminist scholars expose the scientific positivist and androcentric bases for scientific objectivity. The studies from these scholars show the need for a redefinition of technical communication.

Moreover, in the 1980s and 1990s, technical communication has adapted ethnography, an anthropological research method, to explore workplace environments. The most recent ethnographic studies in technical communication parallel the concerns of feminist scholars by acknowledging the subjective point of view of the researcher, looking for messages as well as silences and gaps within communities, and emphasizing group values and lived experience.

The subject of these ethnographic studies in technical communication has often been collaborative writing. Again, feminist theory has much to offer technical writing researchers and teachers in analyzing successful collaboration. Particularly, psychological studies and object-relations theory reveal the familial and cultural roots of women's strong psychological connections with others, and scholars describe the strategies that women frequently use to encourage that closeness. These strategies, if made available to all members of a collaborative writing team, should encourage effective collaboration.

In this essay, I explore how current views of scientific objectivism and the adoption of ethnographic studies—particularly those of collaborative writing—necessitate a new and, perhaps, revolutionary affiliation for technical communication and feminist theory. Although many scholars mentioned in this essay would readily call themselves feminists—those who recognize and wish to correct the unequal treatment of women in our culture—even those who may not feel comfortable with this feminist label have conducted work that exposes sexual inequality. To frame this exploration, I first discuss six common characteristics of feminist theory, as well as three issues that divide feminists. These characteristics and issues from the work of feminist literary theorists, the object-relations area of psychology, and feminist critiques of science will be evident in new definitions of technical writing.

## CHARACTERISTICS OF FEMINIST THEORY

Although feminist theorists resist uniformity of definition and methodology, a survey of their theories reveals six common characteristics:

1. celebration of difference
2. theory activating social change
3. acknowledgment of scholars' backgrounds and values
4. inclusion of women's experiences
5. study of gaps and silences in traditional scholarship
6. new sources of knowledge—perhaps a benefit of the five characteristics above.

Discussing the characteristics of feminist theory is difficult. Feminist theorists are often suspicious of traditional studies within history, literature, psychology, and science, because, if these traditional studies address women at all, they portray them as Woman or Other. Feminist theorists then resist a uniform definition of feminist studies to avoid stereotyping women in their roles as scholars or research subjects. According to de Lauretis, feminist studies must be "absolutely flexible and readjustable, from women's own experience of difference, or our difference from Woman and of the difference among women," and this is a shift to the more complex notion that the female subject is a "site of differences," rather than a subject defined by sexual difference [4]. Thus feminists see any unified image of women as reductive: "Instead, having been constrained and divided by definitions imposed upon us by others, we tend to value autonomy and individual development. Definitions, whether formulated by feminists or not, threaten to divide us" [5, p. 73; 6, p. 9]. Moreover, according to Harding, not only have traditional theories made it "difficult to understand women's participation in social life, or to understand men's activities as gendered, but also traditional theories "systematically exclude the possibility that women could be 'knowers' or *agents of knowledge*" [7, p. 3]. Thus the first characteristic of feminist theory becomes resistance to definition and a celebration of diversity. For many feminists, the insistence on diversity consequently becomes a political or activist stance—the second common characteristic of feminist theory. As Harding stated, in resisting a search for "the one, true story of human experience," feminism may avoid replicating the tendency in the patriarchal theories to police thought by assuming that only the problems of some women are reasonable ones" [8, pp. 284-285]. For many, this insistence on diversity is consistent with the larger women's movement [9, p. 162]. Put simply by Weedon, "Feminism is a politics. It is a politics directed at changing existing power relations between women and men in society" [10, p. 1]. Feminist theorists recognize that change will bring positive aspects to women's lives. More specifically, theorists such as Delmar believe that at the

> very least a feminist is someone who holds that women suffer discrimination because of their sex, that they have specific needs which remain negated and unsatisfied, and that the satisfaction of these needs would require a radical change (some would say a revolution even) in the social, economic and political order [6, p. 8].

Most recently, *standpoint* feminist theorists "attempt to move us toward the ideal world by legitimating and empowering the 'subjugated knowledge' of women" [8, pp. 295-296]. Thus the second characteristic of feminist theory is the assumption that new knowledge about women's lives will change and improve those lives—the personal is political.

If feminist criticism and theory are political, the reader should be aware of the feminist writer and feminist critic's beliefs and values. Thus the acknowledgement of scholars' backgrounds and values is the third demand of feminist theory. Moi found this characteristic "one of the fundamental assumptions of any feminist critic to date"; the feminist critic should "supply the reader with all necessary information about the limitations of one's own perspective at the outset" (see also Kaplan) [11, p. 43-44; 12, p. 40]. This third feature of feminism places the researcher on the same plane as the subject, particularly within science. According to Harding, "the class, race, culture, and gender assumptions, beliefs, and behaviors of the researcher her/himself must be placed within the frame of the picture that she/he attempts to paint" [ 7, p. 9]. This admission comes closer to true, rather than simply asserted, objectivity. As Harding stated,

> We need to avoid the 'objectivist' stance that attempts to make the researcher's cultural beliefs and practices invisible while simultaneously skewering the research objects beliefs and practices to the display board . . . . Introducing this 'subjective' element into the analysis in fact increases the objectivity of the research and decreases the 'objectivism' which hides this kind of evidence from the public [7, p. 9].

Revealing the characteristics of the researcher not only helps eliminate bias, but also places the researcher on a more equal level with the subject of the study.

A unique appreciation of both audience and subject further motivates feminists to reveal their own beliefs and behaviors. "Feminist research will be *used* by the audience, because it will provide for women explanations of social phenomena that they want and need" [7, p. 8]. To determine these explanations, the subject matter of feminist research is women's experiences—the fourth characteristic among feminist theorists. Feminists see a definite relation between experience and discourse. "One distinctive feature of feminist research, according to Harding," is that it generates its problematics from the perspective of women's experience. It also uses these experiences as a significant indicator of the 'reality' against which hypotheses are tested" [7, p. 7]. The audience of feminist scholarship benefits from linking literature to life, from texts that engage in "nurturing personal growth and raising the individual consciousness" [11, p. 43]. This encouragement to test text against experience also reveals what is missing within other discourses and theories. When we begin inquiries with women's experiences instead of men's," said Harding, "we quickly encounter phenomena (such as emotional labor and the positive aspects of 'relational' personality structures) that were made invisible by the concepts and categories of these theories" [8, p. 284]. Thus feminist critics relate to their audiences by acknowledging their own backgrounds, by investigating experiences that their audiences have, and by inviting their audiences to test feminist investigations against their own experience.

Seeking the gaps or silences within traditional scholarship—the fifth characteristic of feminist theory—relates to this appreciation of women's experiences. Gaps or silences have been examined in two ways: the identity of the missing and the potential nature of the study had the missing been included. Feminist critics who seek the identity of the missing must decide if deconstruction is helpful, particularly Derridean *difference*. According to Meese, the deconstructive critic "seeks to temporalize or negate the stasis of 'difference' as a structure of paired opposites inscribed and reinscribed forever in a fixed power relationship within a closed system" [13, p. 80]. The deconstructive process opens the gaps in the structure to reveal what women's natures might be if not defined in terms of the opposite of men. Feminist critics, whether in science, literature, psychology, or other disciplines, also speculate what their disciplines might have studied and what methods and discoveries might be sanctioned if women had been included in these disciplines.

Had women been empowered as critics, as audiences, and as sources of experience throughout the histories of the disciplines, they might have established new theories of knowledge and reality—a sixth common characteristic of feminist theory. Feminists acknowledge that what constitutes self-image is not just a matter of personal experience, but that image is "interpreted or reconstructed by each of us within the horizon of meanings and knowledges available in their culture at given historical moments" [4, p. 8]. If women's experiences had contributed to those meanings and knowledges, women would have been a source of knowledge, of what culture determines as reality, and of what scholars canonize. Feminists have struggled with the power of the canon. Canonization, said Kolodny, "puts any work beyond questions of establishing its merit and, instead, invites students to offer only increasingly more ingenious readings and interpretations, the purpose of which is to validate the greatness already imputed by canonization" [9, p. 150]. By asserting that women are subjects or sources of knowledge, rather than objects of study as Other or Woman, feminists empower women and change definitions of reality or canonized texts. Feminism, stated Delmar, transforms women "from object of knowledge into a subject capable of appropriating knowledge, to effect a passage from the state of subjection to subjecthood" [6, p. 25]. Feminism then can ultimately initiate changes not only in political, social, and economic structures but also in sources of knowledge.

## ISSUES IN FEMINIST THEORY

Before assessing the impact that these six characteristics of feminist theory have on new definitions of technical communication, I must acknowledge three issues of debate among feminists, because technical communication scholars may have to decide where they stand on those issues:

1. Should feminists emphasize similarities or differences among men and women?
2. Should these differences be located in cultural or biological traits?
3. Should these first two issues promote or displace binary opposition?

The first issue of debate involves whether women should emphasize their differences from men or their similarities to men. Should women try to take on traditionally defined masculine traits? For example, should women in the workplace learn the language of power? Or should women celebrate what have been labeled feminine traits? For example, as Rosenthal asked, should women "insist on the 'humaneness' of typical female qualities like compassion, help, nurture, and self-sacrifice—to put these qualities into practice in their professions . . ."? [14, p. 66] Or should both men and women move toward androgyny, a "non-sex-marked humanity"? [14, p. 66] The French feminists, such as Cixous, Kristeva, and Irigaray, have chosen the second option to celebrate those traits that have been labeled feminine traits: Moi said that while "extolling women's right to cherish their specifically female values, the French feminists "reject 'equality' as a covert attempt to force women to become like men" [11, p. 98]. Any application of feminist theory to technical communication will have to struggle with the choice of emphasizing similarities or differences between men and women.

The second issue involves controversy about the origins of differences between men and women. Whether or not theorists decide to emphasize the differences or the similarities between gender traits, they recognize traits as biological, social, or a combination of both. Epstein summarized this issue within her definitions of maximalist and minimalist feminist perspectives [14]. The maximalist holds that there are basic differences between the sexes; some proponents ascribe these differences to biology or to social conditioning, whereas others claim the differences are "lodged in the differing psyches of the sexes by the psychoanalytic processes that create identity" [14, p. 25]. "These scholars," said Epstein, "typically believe that differences are deeply rooted and result in different approaches to the world, in some cases creating a distinctive 'culture' of women" [14, p. 25]. Epstein's minimalist position contends that men and women are essentially similar. Gender differences are "superficial" because they are "socially constructed (and elaborated in the culture through myths, law, and folkways) and kept in place by the way each sex is positioned in the social structure" [14, p. 25]. The origin of difference affects the ease with which men or women can assume the traditional traits of the other gender—another issue of importance to the technical communication researcher.

The third issue is whether scholars can and should avoid reinforcing binary opposition in discussions of difference. Moi identified the goal of feminism to "deconstruct the death-dealing binary opposition of masculinity and femininity," for here the feminine is the negative of the masculine, always lower in the hierarchy [11, p. 13]. Feminists such as Epstein warn that by celebrating a woman's culture, feminists may reinforce the opposition and this hierarchy [14, p. 25]. On the other hand, the French feminists emphasize the differences between men and women; for example, they celebrate *jouissance,* a type of sexual and physical joy that can be experienced *only* by women, or as Jones defined it, "the direct re-experience of the physical pleasures of infancy and of later sexuality, repressed but not obliterated by the Law of the Father" [15, p. 87].

These three issues of debate among feminists are highly related. Although feminist theories promote difference and resist uniform definition of feminist methodology, should they promote difference or stress similarity of experience when studying men and women? Moreover, where should the origin of difference be located, within biology or society? Finally, will the result of exploring difference promote or displace binary opposition? Again, these issues cannot be ignored when applying feminist theory to technical communication.

## REDEFINITION OF TECHNICAL COMMUNICATION

These traits and issues of debate within feminist theory inform the pressure to redefine technical communication that comes from exposing the myth of scientific objectivity, adapting ethnographic research techniques, and studying collaborative writing.

## The Myth of Objectivity

Traditional definitions of technical communication affiliate it with the quantitative and objective scientific method, calling technical communication a "data retrieval method" for the specialized audience [16, p. 137]. Redefinitions of technical communication have been influenced by composition scholars who question the classical distinction made between rhetoric and science. For example, Berlin proposed a "New Rhetoric," which acknowledges that truth is "dynamic and dialectic," and that language "creates the 'real world' by organizing it, by determining what will be perceived and not perceived, [and] by indicating what has meaning and what is meaningless" [17, pp. 774-775]. He opposed the popular Current-Traditional Rhetoric, with its link to scientific positivism and the myth of objective reality—the assumption that truth could be discovered through the experimental or scientific method if the individual was "freed from the biases of language, society, or history" [17, p. 770]. Berlin, in some sense, defied the Aristotelian binary opposition of rhetoric and science.

Thus technical communication scholars such as Halloran, relating to Kuhn, admit that "in a very fundamental way," science is "argument among scientists," but that technical communication maintains the deceptive ethos of the "dispassionate, disinterested truth-seeker" [18, p. 85]. Miller asserted that technical communication, as commonly taught, is "shot through with positivist assumptions" [19, p. 613]; instead, technical communication should be viewed as a matter of "conduct rather than of production, as a matter of arguing in a prudent way toward the good of the community rather than of constructing texts" [20, p. 23]. Samuels, reacting to Halloran, Miller, as well as Kuhn, then defined technical communication as a "recreation of reality for special purposes"; rather than transmitting or inventing reality, the communicator "extends" perceptions of truth [21, p. 11]. Finally, Dobrin decided that technical communication simply "accommodates technology to the user," for any claim of objectivity is based on scientific domination (see also Goldstein) [22, p. 242; 23, p. 25]. As the distinctions between science and rhetoric disappear, truth is defined as agreement within a community, not as discoverable and describable reality. Technical communication then offers culturally based perceptions to the audience, rather than objective information and data.

Although gender roles are part of culture, few scholars, so far, have examined the impact of these roles on the technical communicator. Sterkel did a quantitative study that suggests that women have adopted the language of power in business writing [22]. Smeltzer and Werbel found that the type of communication required makes more difference than does the gender of the writer, and Tebeaux discovered distinct differences among male and female inexperienced business

writers [23, 24]. However, if new definitions of technical communication acknowledge the culturally based perceptions within scientific and technical discourse,
gender studies of science and technology must change the way technical communication scholars view their field.

Over the last decade, feminist scholars have identified masculine bias within the
discovery and discourse of science and technology. These biases, according to
Bleier, are both the source of science's "great strength and value" and of its
"oppressive power" [25, p. 57]. For example, in Keller's examination of the
genderization of science, she traced the mythology that assigns objectivity,
reason, and mind to the male, and subjectivity, feeling, and nature to the female
[26, pp. 6-7]. This binary opposition limits women's experiences as valid scientific subjects, as well as prevents women from being sources of scientific
knowledge. Keller concluded that in the family structure masculinity is associated
with "autonomy, separation, and distance":

> Thus it is that for all of us—male and female alike—our earliest experiences incline
> us to associate the affective and cognitive posture of objectification with the mas-
> culine, while all processes that involve a blurring of the boundary between subject
> and object tend to be associated with feminine [26, p. 87].

Keller's study of scientist Barbara McClintock's disregard for the traditional
separation from her subject reveals a gender-free approach to science [27, p. xvii].
Rather than using a static objectivity that distances scientists from their subjects,
Keller proposed a dynamic objectivity that "actively draws on the commonality
between mind and nature as a resource for understanding" [26, p. 117]. Keller,
Bleier, and other feminist theorists expose the biases of the scientist hidden behind
the ethos of objectivity and identify women's experiences within the gaps and
silences of traditional science.

The resistance toward including women's experiences and employing women
as sources of knowledge is particularly strong in science. Perhaps for this reason,
acknowledging the connections between feminist theory and technical communication will be difficult for many. Within science and central to the image of
masculinity, according to Harding, is the "rejection of everything that is defined
by culture as feminine and its legitimated control of whatever counts as feminine"
[28, p. 54]. In a sense, science has identified the masculine with the human and so
has excluded the feminine. Also, science has tended to define femininity by
biological, rather than cultural, traits. In particular, a woman's reproductive
capacity is seen by science as "an immense biological burden, condemning her to
the world of nature, of the body, of emotions, and subjectivity" [29, p. 44]. And so,
many eminent scientists have concluded, as did nuclear physicist Rabi in 1982,
that women are "temperamentally unsuited to science" because of their nervous
systems:

It makes it impossible or them [women] to stay with the thing. I'm afraid there's no use quarrelling with it, that's the way it is. Women may go into science, and they will do well enough, but they will ever do great science [in 29, p. 36].

This resistance to women's concerns carries over into technology and the social sciences. For example, Hacker found that engineers often described social sciences as womanly—"soft, inaccurate, lacking in rigor, unpredictable, amorphous" [30, p. 345]. She also found that engineers assign more status to areas that seemed the "cleanest, hardest, most scientific" such as electrical engineering and less status to such fields as civil engineering that were more involved in social sciences [30, p. 345]. In turn, the social sciences, in particular psychology, often assign higher ranking to experimentalists, seen as particularly objective and linked to *hard* science, and lower ranking to developmentalists who might have more *social* concerns. As with engineering, areas in psychology that could take on the appearance of objectivity appear at the top of the hierarchy [31, pp. 41-42].

In affiliating with scientific positivism and in defining itself as the objective transfer of data, truth, and reality, traditionally defined technical communication ranks higher than other supposedly *subjective* types of writing, engages in dualistic thinking, and maintains closeness with patriarchal institutions of power. Therefore, to enhance legitimacy for their field, technical communication scholars and teachers may resist redefinition that divorces technical writing from this source of power. However, feminist theorists affect the recent redefinitions of technical communication as made by scholars such as Miller, Samuels, Dobrin, and Halloran. Feminist theorists challenge technical communicators to reevaluate their fields by exposing the masculine bias of science and technology, insisting that women cease to be the object and instead become the subject of science, and defying the dualism of masculine/feminine, objective/subjective, and culture/nature. This revision, I believe, will be most useful as technical communication scholars employ ethnographic studies, particularly those that study collaborative writing.

## Ethnographic Studies

With the current emphasis on both the social nature of writing and the ways in which a discourse community produces documents, ethnographic methods have been adopted by technical communication scholars. To use ethnographic methods, technical communicators must study what constitutes a discourse community, what and why interactions take place within that community, what texts are produced, what subjects are considered appropriate within those texts, how genres are evolved, and how methods of inquiry are chosen and approved [32, p. 241]. Ethnography is also appropriate because of the interdisciplinary nature of technical communication:

> Because those of us teaching business and technical communication possess a wide range of disciplinary training—from linguistics to literature to business education to computerized-document design—we can bring a multidisciplinary perspective to ethnographic research [33, p. 30].

The cultures that technical communication ethnographers study within the industrial setting and their respect for subjects "who are, in some ways, far more expert and knowledgeable than are the ethnographers" are most essential to this ethnographic research [34, p. 507]; in technical communication ethnographies, there is no Other. The technical communication audience tests what the ethnographer says against subjects' own experiences.

Parallels between ethnography and feminism include multiplicity, acknowledgment of the researcher's values and background, appreciation of lived experience and new sources of knowledge, and discovery—rather than testing—of meaning. Kantor's five characteristics of communication ethnography include these parallels: (a) contextuality, (b) researcher as participant-observer, (c) multiple perspectives, (d) hypothesis generating, and (e) meaning making [35, pp. 72-74]. Because ethnography stresses that behavior is expressed and influenced by the groups and the cultures to which individuals belong, the ethnographer spends long periods of time within a community to get detailed, concrete records of that community's behavior, including language and communication. The ethnographer's perceptions become part of the record: "Typically researchers begin by assessing their own knowledge, experiences, and biases, and reevaluate those influences as their study proceeds" [35, p. 73]. This stance of being both participant and observer can be called "disciplined subjectivity" [35, p. 73].

Ethnographers use triangulation or more than one means of record keeping—sometimes a combination of interviews, field notes, and diaries—and seek the reactions from other researchers or community members to enhance their interpretations. Within their observations ethnographers generate rather than test hypotheses; they may develop more research questions than they answer. The purpose of ethnography is "to look at ways in which individuals construct their own realities and shared meanings" [36, p. 298]. In ethnographic thick description, the researcher, much like the feminist scholar, records the daily details of community life. The ethnographer then discovers within this detail the meanings and values that people, not just those with power, attribute to phenomena. By triangulation, the ethnographer seeks multiple impressions; the feminist theorist in turn finds multiplicity essential in integrating women, not Woman, into the world picture. Much like feminist theorists, the ethnographic observer-participants examine and admit their own background and cultural bias, including their gender roles within the observed community and the audience of the ethnographic description. Ethnographers share with feminist theorists the goal of understanding rather than evaluating a community, seeking new meanings within previous gaps and silences, and finding new sources for that knowledge. "Ethnography attempts to bring stories not yet heard to the attention of the academy," concluded Brodkey [37, p. 48].

In addition to including women's lives in their studies, female ethnographers have speculated recently about the ways their sex affects their assimilation into a community and their consequent research. Again, ethnographic narratives in general "jeopardize the positivist campaign to deny anyone's lived experience, in the name of objectivity" [37, p. 41]. However, female ethnographers face recurring issues, such as protective behavior triggered by their sex and the great difference between their own life-styles and those of the women they study [38-40].

Feminist traits are inherent in contemporary ethnographic methodology. Ethnographers reject the received view within social science, as Agar defined it, "a view that centers on the systematic test of explicit hypotheses"; ethnography does not claim that anyone using the same methods would come to the same conclusions [41, p. 11]. At the least, ethnographers' varied backgrounds and intended audiences cause different conclusions. The traditional scientific hierarchy between researcher and subject is abolished, and connections are sought:

> Ethnographers set out to show how social action in one world makes sense from the point of view of another. Such work requires an intensive personal involvement, an abandonment of traditional scientific control, an improvisational style to meet situations not of the researcher's making, and an ability to learn from a long series of mistakes [41, p. 12].

Rather than seeking similarities or universals as in traditional science, the ethnographer reacts to breakdowns or differences. These breakdowns are resolved "by changing the knowledge in the ethnographer's tradition," or as in feminist theory, sources of knowledge are not dictated by the power elite [41, p. 25]. In this way, ethnography has activist characteristics.

In their challenges to the received view within social science, ethnographers question binary opposition or dualistic thinking about qualitative and quantitative research. Firestone characterized ethnographers as pragmatists—as opposed to purists—and believed that ethnographers contrast quantitative researchers who assume that there are "social facts with an objective reality apart from the beliefs of individuals" to qualitative researchers who believe that "reality is socially constructed through individuals or collective *definitions of the situation* (see also Howe) [42, pp. 16, 43]. In fact, pragmatists like Goetz and LeCompte suggested that ethnography is more objective than quantitative research because ethnographers admit their subjective experiences (see also Hymes; Geertz; and North)— a conclusion identical to Harding's [43, p. 9; 44-46]. Ethnography, then, is rhetorical because ethnographers must understand and influence their audiences (see Kleine) [47]. Thus the ethnographer, again like the feminist theorist, attempts to incorporate into the canon research methods and subjects that were excluded by scientific positivism and a quantitative focus, and the ethnographer questions the binary opposition that excluded these research methods and subjects in the first place.

More particularly, the technical communication ethnographer frequently studies how editors, writers, technical developers, potential customers, and graphic artists collaborate to produce a document. Composition specialists, such as Bruffee, have for over a decade stressed the social nature of writing and questioned the image of the solitary writer. According to Bruffee, texts are "constructs generated by communities of like-minded peers" [48, p. 774]. Some composition researchers, in particular LeFevre, assert that the myth of the solitary writer complements a gender-biased social view: "The persistence of such an ideal of individual autonomy in male-centered, capitalistic culture further explains why a Platonic view of invention, which stresses the writer as an isolated unit apart from material and social forces, has been widely accepted" [49, p. 22]. Therefore, technical communication researchers must attend to how gender roles affect industrial collaborative writing.

Ethnographic studies of the workplace reveal that effective collaborators have good interpersonal skills, the ability to connect and maintain connections with collaborators even in times of conflict over ideas (see, for example, Doheny-Farina or Debs) [50, p. 181; 51, p. 3]. Researchers do stress that collaborators "should be reassured that conflict over ideas, over substantive matters, can be a positive development in the collaborative process" [52, p. 121]. In addition, feminist scholars Keller and Moglen proposed, "Under certain circumstances, cooperation may actually be facilitated by differentiation and autonomy" [53, p. 27]. However, this substantive conflict is most productive in solid relationships in which people have developed enough trust to self-disclose and to not feel threatened by criticism. Thus Ede and Lunsford profiled effective collaborators as flexible, respectful of others; attentive and analytical listeners; able to speak and write clearly and articulately; dependable and able to meet deadlines; able to designate and share responsibility, to lead and to follow; open to criticism but confident in their own abilities; ready to engage in creative conflict [54, p. 66]. Feminist theorists, particularly object-relations theorists in psychology, enhance these ethnographic studies of collaborative writing, because these theorists relate how and why males and females connect and cooperate.

## Collaborative Writing

Studies reveal that girls and boys in the pre-Oedipal stage of life differ in their urge to connect with or distinguish themselves from others. Because mothers still do so much of the parenting, girls identify with this main parenting figure, whereas boys identify with the father who appears independent and involved in the world outside the home. As Chodorow concluded,

> Girls emerge from this period with a basis for 'empathy' built into their primary definition of self in a way that boys do not. Girls emerge with a stronger basis for experiencing another's need or feeling as one's own (or of thinking that one is so experiencing another's needs and feelings) [55, p. 167].

Girls define themselves as "continuous with others," whereas boys define themselves as "more separate and distinct": "The basic feminine sense of self is connected with the world, the basic masculine sense of self is separate" [55, p. 169]. Boys may develop into men who avoid close connections because these connections threaten their sense of self, whereas girls may develop into women who seek close relationships and assume responsibility for cooperation.

Gilligan, in her study of adult ethics, stated that men may see connections as threatening to their place in the competitive work hierarchy, whereas women fear being "too far out on the edge"—too far from connections with others [1, p. 43]. Building on the research of Chodorow and Gilligan, Belenky and her coauthors distinguished the separate knowing of many men from the connected knowing of many women. In groups, authority or knowing for women is commonality of experience, which requires "intimacy and equality between self and object, not distance and impersonality" [56, p. 183]. Obviously, gender roles, as established in the family structure, influence how men and women relate to the demands of collaboration—a topic of much interest to technical communicators.

Feminist theorists' elevation of women's experiences and knowledge and the feminist debate over the origin of difference inform studies of collaboration. If women more easily *connect*—a virtue when functioning on a writing team—can men learn these connecting strategies too? Minimalists, as defined by feminist theorist Epstein, would see that change as quite possible, although usually feminists have studied the ease with which women have taken on traditional male traits. Because gender roles are socially constructed—created by nurture rather than nature—they can be changed. According to Flynn,

> Women share interpretive frameworks and strategies because they have had common experiences, ones different, for the most part, than those of men. Such a position is optimistic in the sense that it posits that different experiences can produce different interpretations [57, p. 4].

Thus gender roles are not static.

Whether optimistic about this change or not, feminist theorists at the very least can point out *what* makes it difficult for men to balance competitive impulses with cooperative needs. Again, object-relations theorists stress basic differences in masculine and feminine self identity and the difficulty in overcoming the effects of the family structure: "The boy's repression of the female aspect of himself is one of the reasons men find it hard to be nurturant as adults" [58, p. 178]. And feminist theorists, who identify themselves as practicalists—or believe that thinking arises from practices—suggest that "maternal thinking" comes from maternal practice and is accessible to all who practice it [59, pp. 13-14]. Knowing why men resist connections and anticipating the effects of men's increasing participation in parenting are the first steps to overcoming this resistance.

Because of these gender roles, men's and women's attitudes toward conflict during collaboration differ. Technical communicators need to learn what feminist theorists say about these different gender roles in order to manage conflict effectively, which helps to ensure the success of collaboration. Women are taught to avoid conflict and may view all conflict as interpersonal and potentially damaging to relationships [60, pp. 20-22]. Men tend to view conflict as healthy competition and as primarily substantive [61, p. 61]. In resolving conflict, men and women also use different strategies: "Two males in a conflict typically employ bargaining techniques, logical arguments, and anger to manage the situation. In contrast, two females in a conflict situation focus on understanding each other's feelings" [62, pp. 47-48]. Again, understanding the cultural and familial origins of these differences through feminist studies is the first step toward effective collaboration. Finally, feminist theorists can help technical communicators provide new models of effective collaboration—models that help collaborators break out of gender roles. These new models would also help collaborators value women's experiences and strategies, because these strategies include interpersonal skills. Such strategies—well documented in communication and feminist theory—include self-disclosure, sensitivity to nonverbal cues, perception of others' emotions, questioning intonations in responses, acknowledgment of previous speakers, and resolution of conflict in nonpublic ways [63, p. 275; 64, p. 192; 65, p. 10; 66, p. 120].

One of the most recent new models of collaborative writing comes from Ede and Lunsford, who recognized that a dialogic model is articulated mainly by the female technical writers they interviewed. Ede and Lunsford described the traditional model of collaboration as hierarchal, with productivity and efficiency the goal. In the traditional model, "the realities of multiple voices and shifting *authority* are seen as difficulties to be overcome or resolved" [54, p. 133]. On the other hand, in the dialogic model, roles are fluid, each collaborator may take on "multiple and shifting roles," and "the process of articulating goals is often as important as the goals themselves and sometimes even more important" [54, p. 133]. Ede and Lunsford also said, "Furthermore, those participating in dialogic collaboration generally value the creative tension inherent in multivoiced and multivalent ventures" [54, p. 133]. This dialogic model is predominantly feminine, "So clearly 'other,'" and Ede and Lunsford lamented that there is no "ready language" to describe it [54, p. 133]. However, feminist theorists could supply that language to Ede and Lunsford; Jordan and Surrey label the capacity to move from one perspective to another as "oscillating self-structure" [67, p. 92]. Although Jordan and Surrey applied the term to the mothering process, the term well represents the process that Ede and Lunsford observe in dialogic collaboration. As technical communication scholars explore the collaborative writing process, they can learn much from feminist studies.

Linking technical communication with feminist theory may seem alien to many. However, at the very least, the interdisciplinary nature of technical communication will lead the field in the direction of feminist theory. As feminist theorists attack the last vestiges of scientific positivism within science and technology, technical communication must also let go of the ethos of the objective technical writer who simply transfers information and accept that writers' values, background, and gender influence the communication produced. As technical communicators convince their audiences to accept a version of reality, they develop persuasion strategies by identifying with their audiences. In many major corporations, technical writers are the customers' representatives and advocates and must discover the fine details of their customers' experiences.

As technical communicators explore collaborative writing through ethnographies, they again adopt stances similar to feminist theorists. Their own backgrounds and values as ethnographers must be admitted to their audiences. They must seek the many voices of those who witness and experience the culture they investigate. These technical communicators also expose the gaps and silences in previous studies and identify new sources of knowledge that may challenge dominant or traditional cultures.

The mission of most technical communication scholars is to prepare future technical writers to enter industry and to improve the industrial processes that produce communications. Because so many documents are collaboratively produced and their effectiveness threatened when collaborative teams suffer interpersonal conflicts, again the work of feminist theorists can show how these conflicts may be affected by gender roles. New models of collaboration, such as those suggested by Ede and Lunsford, recognize the effects of collaborators' gender roles.

In acknowledging these connections with feminist theory, technical communication must wrestle with the issues that confront feminist theorists. In suggesting effective collaborative strategies, should technical communicators stress the similarities or differences between men and women? In doing so, can they avoid the labeling that contributes to dualistic thinking or binary opposition? Should male collaborators know that they are being encouraged to adopt what have been labeled *female* interpersonal traits? Are the sources of these gender traits primarily social or biological? If they are social, collaborators can move more easily toward androgyny. If not, change will be more difficult. How will or how should the industrial setting be affected by these ethnographic studies of gender roles and collaborative traits? Should technical communicators become activists in industrial settings as they explore gender roles within collaboration? The answers to these questions remain for future technical communicators to discover. However, technical communication must be redefined to include these issues.

## REFERENCES

1. C. Gilligan, *In a Different Voice: Psychological Theory and Women's Development*, Harvard University Press, Cambridge, Massachusetts, 1982.
2. M. F. Belenky et al., *Women's Ways of Knowing: The Development of Self, Voice,* and *Mind,* Basic Books, New York, 1986.
3. S. Helgesen, *The Female Advantage: Women's Ways of Leadership*, Doubleday, New York, 1990.
4. T. de Lauretis, Feminist Studies/Critical Studies: Issues, Terms, and Contexts, in *Feminist Studies/Critical Studies*, T. de Lauretis (ed.), Indiana University Press, Bloomington, Indiana, 1986.
5. E. Meese, *Crossing the Double-Cross: The Practice of Feminist Criticism*, University of North Carolina Press, Chapel Hill, North Carolina, 1986.
6. R. Delmar, What Is Feminism? J. Mitchell, in *What is Feminism: A Re-Examination*, J. Mitchell and A. Oakley (eds.), Pantheon, New York, 1986.
7. S. Harding, Is There a Feminist Method?, in *Feminism and Methodology*, S. Harding (ed.), Indiana University Press, Bloomington, Indiana, 1987.
8. S. Harding, The Instability of the Analytical Categories of Feminist Theory, in *Sex and Scientific Inquiry*, S. Harding and J. F. O'Barr (eds.), University of Chicago Press, Chicago, 1987.
9. A. Kolodny, Dancing through the Minefield: Some Observations on the Theory Practice, and Politics of a Feminist Literary Criticism, in *The New Feminist Criticism: Essays on Women, Literature, and Theory*, E. Showalter (ed.), Pantheon, New York, 1985.
10. C. Weedon, *Feminist Practice and Poststructuralist Theory*, Basil Blackwell, New York, 1987.
11. T. Moi, *Sexual/Textual Politics: Feminist Literary Theory*, Metheun, New York, 1985.
12. S. J. Kaplan, Varieties of Feminist Criticism, in *Making a Difference: Feminist Literary Criticism*, G. Greene and C. Kahn (eds.), Metheun, New York, 1985.
13. E. Meese, *Crossing the Double-Cross: The Practice of Feminist Criticism*, University of North Carolina Press, Chapel Hill, North Carolina, 1986.
14. C. F. Epstein, *Deceptive Distinctions: Sex, Gender, and the Social Order*, Yale University Press, New Haven, Connecticut, Russell Sage, New York, 1988.
15. A. R. Jones, Writing the Body: Toward an Understanding of L'Ecriture Feminine, in *Feminist Criticism and Social Change: Sex, Class, and Race in Literature and Culture*, J. Newton and D. Rosenfelt (eds.), Metheun, New York, 1985.
16. J. S. Harris, On Expanding the Definition of Technical Writing, *Journal of Technical Writing and Communication,* 8:2, pp. 133-138, 1978.

17. J. A. Berlin, Contemporary Composition: The Major Pedagogical Theories, *College English, 44*:8, pp. 765-777, 1982.
18. M. S. Halloran, Technical Writing and the Rhetoric of Science, *Journal of Technical Writing and Communication, 8*:2, pp. 77-88, 1978.
19. C. Miller, A Humanistic Rationale for Technical Writing, *College English, 40*:6, pp. 610-617, 1979.
20. C. Miller, What's Practical About Technical Writing?, in *Technical Writing: Theory and Practice*, B. E. Fearing and W. K. Sparrow (eds.), Modern Language Association, New York, 1989.
21. M. S. Samuels, Technical Writing and the Recreation of Reality, *Journal of Technical Writing and Communication, 15*:1, pp. 3-13, 1985.
22. K. S. Sterkel, The Relationship between Gender and Writing Style in Business Communication, *Journal of Business Communication, 25*:4, pp. 17-38, 1988.
23. L. R. Smeltzer and J. D. Werbel, Gender Differences in Managerial Communication: Fact or Folk-Linguistics?, *Journal of Business Communication, 23*:2, pp. 41-50, 1986.
24. E. Tebeaux, Toward an Understanding of Gender Differences in Written Business Communication: A Suggested Perspective for Future Research, *Journal of Business and Technical Communication, 4*:1, pp. 23-43, 1990.
25. R. Bleier, Lab Coat: Robe of Innocence or Klansman's Sheet, in *Feminist Studies/Critical Studies*, T. de Lauretis (ed.), Indiana University Press, Bloomington, Indiana, 1986.
26. E. F. Keller, *Reflections on Gender and Science*, Yale University Press, New Haven, Connecticut, 1985.
27. E. F. Keller, *A Feeling for the Organism*, Freeman, San Francisco, California, 1983.
28. S. Harding, *The Science Question in Feminism*, Cornell University Press, Ithaca, New York, 1986.
29. V. Gornick, *Women in Science: 100 Journeys into the Territory*, Simon & Schuster, New York, 1990.
30. S. L. Jacker, The Culture of Engineering: Woman, Workplace, and the Machine, in *Women, Technology, and Innovation*, J. Rothschild (ed.), Pergamon, New York, 1982.
31. C. W. Sherif, Bias in Psychology, in *Feminism and Methodology*, S. Harding (ed.), Indiana University Press, Bloomington, Indiana, 1987.
32. L. Faigley, Lester. Nonacademic Writing: The Social Perspective, in *Writing in Nonacademic Settings*, L. Odell and D. Goswami (ed.), Guilford, New York, 1985.
33. J. W. Halpern, Getting in Deep: Using Qualitative Research in Business and Technical Communication, *Journal of Business and Technical Communication, 2*:2, pp. 22-43, 1988.
34. S. Doheny-Farina and L. Odell, Ethnographic Research on Writing: Assumptions and Methodology, in *Writing in Nonacademic Settings*, L. Odell and D. Goswami (eds.), Guidlford, New York, 1985.
35. K. J. Kantor, Classroom Contexts and the Development of Writing Intuitions: An Ethonographic Case Study, in *New Directions in Composition Research*, R. Beach and L. S. Bridwell (eds.), Guilford, New York, 1984.
36. K. J. Kantor, D. R. Kirby, and J. P. Goetz, Research in Context: Ethnographic Studies in English Education, *Research in the Teaching of English, 15*:4, pp. 293-309, 1981.

37. L. Brodkey, Writing Ethnographic Narratives, *Written Communication, 4*:1, pp. 25-50, 1987.
38. P. Golde, *Women in the Field: Anthropological Experiences*, (2nd Edition), University of California Press, Berkeley, California, 1986.
39. C. A. B. Warren, *Gender Issues in Field Research*, Sage, Beverly Hills, California, 1988.
40. T. L. Whitehead and M. E. Conaway, *Self, Sex and Gender in Cross-Cultural Fieldwork*, University of Illinois, Urbana, Illinois, 1986.
41. M. H. Agar, *Speaking of Ethnography*, Sage, Beverly Hills, California, 1986.
42. W. A. Firestone, Meaning in Method: The Rhetoric of Quantitative and Qualitative Research, *Educational Researcher, 16*:7, pp. 16-21, 1987.
43. J. P. Goetz and M. D. LeCompte, *Ethnography and Qualitative Design in Educational Research*, Academic Press, New York, 1984.
44. D. Hymes, *Language in Education: Ethnolinguistic Essays*, Center for Applied Linguistics, Washington, D.C., 1980.
45. C. Geertz, *The Interpretation of Cultures: Selected Essays*, Basic Books, New York, 1973.
46. S. M. North, *The Making of Knowledge in Composition: Portrait of an Emerging Field*, Boynton Cook, Upper Montclair, New Jersey, 1987.
47. M. Kleine, Beyond Triangulation: Ethnography, Writing, and Rhetoric, *Journal of Advanced Composition, 10*:1, pp. 117-125, 1990.
48. K. A. Bruffee, Social Construction, Language, and the Authority of Knowledge: A Bibliographic Essay, *College English, 48*:8, pp. 773-790, 1986.
49. K. B. LeFevre, *Invention as a Social Act*, Southern Illinois University Press, Carbondale, Illinois, 1987.
50. S. Doheny-Farina, Writing in an Emerging Organization: An Ethnographic Study, *Written Communication, 3*:2, pp. 158-185, 1986.
51. M. E. Debs, Collaborative Writing: A Study of Technical Writing in the Computer Industry, *Dissertation Abstracts International, 47*, Rensselaer Polytechnic Institute, 1986.
52. B. Karis, Conflict in Collaboration: A Burkean Perspective, *Rhetoric Review, 8*:1, pp. 113-126, 1989.
53. E. F. Keller and H. Moglen, Competition: A Problem for Academic Women, in *Competition: A Feminist Taboo?* V. Miner and H. E. Longino (eds.), Feminist Press, New York, 1987.
54. L. Ede and A. Lunsford, *Singular Texts/Plural Authors: Perspectives on Collaborative Writing*, Southern Illinois University Press, Carbondale, Illinois, 1990.
55. N. Chodorow, *The Reproduction of Mothering: Psychoanalysis and the Sociology of Gender*, University of California Press, Berkeley, California, 1978.
56. M. F. Belenky et al., *Women's Ways of Knowing: The Development of Self, Voice, and Mind*, Basic Books, New York, 1986.
57. E. A. Flynn, *Toward a Feminist Social Constructionist Theory of Composition*, Conference on College Composition and Communication, Chicago, March 1990.
58. J. Flax, The Conflict Between Nurturance and Autonomy in Mother-Daughter Relationships and Within Feminism, *Feminist Studies, 4*:2, pp. 171-189, 1978.

59. S. Ruddick, *Maternal Thinking: Toward a Politics of Peace*, Beacon, Boston, 1989.
60. M. M. Lay, Interpersonal Conflict in Collaborative Writing: What We Can Learn from Gender Studies, *Journal of Business and Technical Communication, 3*:2, pp. 5-28, 1989.
61. J. L. Hocker and W. M. Wilmot, *Interpersonal Conflict*, Brown Dubuque, Iowa, 1985.
62. L. L. Putnam, Lady You're Trapped: Breaking Out of Conflict Cycles, in *Women in Organizations: Barriers and Breakthroughs*, J. P. Pilotta (ed.), Waveland, Prospect Heights, Illinois, 1983.
63. M. L. Knapp, D. G. Ellis, and B. A. Williams, Perceptions of Communication Behavior Associated with Relationship Terms, *Communication Monographs, 47*:4, pp. 262-278, 1980.
64. J. E. Baird, Jr., Sex Differences in Group Communication: A Review of Relevant Research, *Quarterly Journal of Speech, 62*:1, pp. 179-192, 1976.
65. R. A. Hall and B. Sandler, *The Classroom Climate: A Chilly One for Women?* Project on the Status and Education of Women, Association of American Colleges, Washington, D.C., 1982.
66. P. A. Treichler and C. Kramarae, Women's Talk in the Ivory Tower, *Communication Quarterly, 31*:2, pp. 118-132, 1983.
67. J. V. Jordan and J. L. Surrey, The Self-in-Relation: Empathy and the Mother-Daughter Relationship, in *The Psychology of Today's Woman: New Psychoanalytic Visions*, T. Bernay and D. W. Cantor (eds.), Harvard University Press, Cambr Cambridge, Massachusetts, 1986.

# GENDER ISSUES IN TECHNICAL COMMUNICATION STUDIES: AN OVERVIEW OF THE IMPLICATIONS FOR THE PROFESSION, RESEARCH, AND PEDAGOGY

## *Jo Allen*

One of the primary truths of late twentieth-century economics, philosophy, and morality, especially in American culture, is that we absolutely must take advantage of all the intellect, productivity, and creativity at our disposal. One resource that has been traditionally underused or even ignored, is women. Profit-motivated corporations led by progressive, realistic thinkers are recognizing the value—and the urgency—of recruiting, retaining, and promoting women workers (see Mandall and Kohler-Gray) [1, 2]. As more opportunities become open to women—although this number of opportunities is still unsatisfactory—we must encourage women to seek educations that will prepare them for the kinds of careers that will be needed to run our businesses in the twenty-first century. The rewards of these educations extend, of course, far beyond increasing the value of women to American businesses; we have historically associated education with better-paying jobs, more fulfilled lives, greater opportunities and options in life, and higher standards of living.

In recognition of changes for women in business and other facets of contemporary life, the past thirty years have granted prominent attention to gender issues in almost every area of the academic curriculum. Thus, although our educational system has sought ways to entice and accommodate women students, women have themselves become the subject of some of the changes in the curriculum. Studies in current events, politics, and history have charted the initiation and effects of the various phases of the women's movement. Studies in business and economics reflect the effects of the growing numbers of women in the workplace; these studies are especially interested in the effects of women in management. Studies in literature demonstrate a new awareness of and appreciation for the contributions of women writers. Psychology and sociology studies acknowledge the potential influences of gender differences on writing beyond their manifestations in literary pursuits.

Although Mary Lay wrote an exceptionally strong argument for integrating what we have learned from gender studies into our research and applications of collaborative writing, other issues of gender have found virtually no place in our studies—at least they have not shown up in our literature in any significant way beyond cursory references and an occasional article on unbiased language usage (see, for example, Christian; Martyna; Corbett; Vaughn; Wilcoxon)

[3-8]. I suspect that at the heart of our failures to address gender issues in communication is a discomfort with the kinds of research and assumptions generated by questions about gender differences. In our fears of being perceived as sexist, we have simply ignored a vast potential for research. Further, however, we have failed to undertake an important facet of industry-watching; we have failed to study our own industry for the effects of a more diverse work group, one composed of more women than in previous decades. And we have failed to incorporate ideas and the acknowledgements of that diverse work group in our classrooms.

If we concede that the workplace is, indeed, becoming more globally affected and domestically integrated, we should also recognize that this integration influences and changes that workplace. Management theories, personnel research, and organizational studies frequently try to describe and accommodate these changes, but no one has considered the comprehensive effects of these changes on technical communication. Of course, a comprehensive analysis is impractical for a single article, but it is certainly worthwhile to highlight the kinds of changes we practitioners, researchers, and teachers should anticipate in technical-communication studies. In this article, therefore, I present an overview of some of these changes and their potential effects. I have limited this work to concentrate on gender-related issues, but I certainly do not mean to exclude the potential for similar questions and research on race and ethnicity as well. At the end of the article, I suggest some conclusions about the future of our field as it recognizes and accommodates issues of gender within American corporations, research, and classrooms.

## THE TECHNICAL WRITING PROFESSION

Some elements of the technical writing industry have been documented in surveys conducted by the Society for Technical Communication (STC). Although the instrument shifts from survey year to survey year—a real problem when trying to chart comprehensive trends—Russell Stoner compared these surveys, factoring in their shifts in survey instruments to show trends in the technical communication industry [9]. For our purposes, the comparisons of hiring trends and salary trends are most significant over the fifteen-year period that these surveys encompass. The increase in the number of female technical communicators over this fifteen-year period—54 percent women in 1985, compared to 20 percent women in 1970—suggests two conclusions:

1. Females are still being socialized and educated to accept and prefer writing (English studies) over math and science, although this preference is evidently expanding to include writing about science and technology (see Daly, Bell, and Korinek; Holden; Marshall and Smith; Berliner; Clement; White) [10-15].

2. Women are, indeed, moving into traditional male occupations.

Most notably, however, the surveys show that although more women are working as technical communicators, their salaries remain less than their male counterparts' salaries. In 1985, the average salary was $27,500 for women, compared to $34,000 for men in the technical communication field. Of course, several factors may influence salaries, but Stoner showed that for the fifteen year period covered by STC surveys, "STC salaries declined in relationship to the [Gross National Product and the Consumer Price Index]" [9, p. 109]. Further, although technical communicators' actual salaries increased over the fifteen year period, the growing feminization of technical communication corresponds to an overall slower rise in the profession's salaries in comparison to salaries in engineering, nursing, teaching, and social work [9, p. 109]. In other words, technical communicator's salaries have not kept pace with the increases in these other four fields. Elaine Sorensen used a bivariate selectivity approach that controls mediating factors—such as whether women work and whether they choose their occupations, as well as factors such as education, work experience, and geographic area—and found that women working in feminized occupations earn "6-15% less than women with the same characteristics working in other occupations" [16, p. 624]. Thus there is a distinct correlation between low salaries and the feminization of a field.

Although Frederick M. O'Hara, Jr. stated that these trends pairing the number of women in technical communication and slower rising salaries "could not account for the effects of length of employment, level of education, and type of employment on median salaries"—and Stoner made the same point in his article—other criteria are frequently used as reasons that women are not promoted or are not paid as well as men [17, p. 311; 9, p. 109]. As economists and sociologists have pointed out, however, "it is far from clear that the age-productivity profile parallels the age-earnings profile" (quoted in Jacobs) [18, p. 173]. In other words, issues such as seniority often become the basis for salary decisions even when we cannot document that senior workers provide more benefits to the company for which they work. Is seniority, therefore, a reliable instrument for determining an employee's value? Or should corporations look for an instrument that rewards productivity instead of seniority? (For research on gender-related hiring trends and salaries, see Sarkis; Coverdill; Bozzi; Olion, Schwab, and Haberfeld; Gerhart; Tienda, Smith, and Ortiz; Fischer; Glick, Zion, and Nelson; Goldin and Polachek) [19-27].

Further, we must consider the results of studies, such as that by Muniz et al., that find that women tend to be more fluent than men in both written and oral communications in-relation to linguistic, semantic, and ideative fluency [28]. Further, studies in education consistently show that female students tend to be better writers than are male students (see, for example, White) [15]. I point out these kinds of studies merely to raise questions about traditional bases for salary and promotion decisions. In determining salaries and promotions for communicators, some research favors women, but traditions—such as valuing seniority—favor men. Neither research nor tradition, however, is a particularly reliable means of determining the value of any individual writer. It seems most logical to assume that the two sides should cancel each other out—and that salary and promotion decisions should be individually determined based on productivity and quality.

Another flaw in O'Hara's assessment that no bias exists within the technical communication industry is noted in the following line of reasoning:

> The same salary survey . . . showed that [Society for Technical Communication] members' salaries increased 11.5% from 1985 to 1988, which is about the same as the rate of inflation for those years. In other words, technical communicators held their own during that period. This increase, it should be noted, came when women's membership in the Society was at its highest, and the rate of salary increase did not decline with this increase in the proportion of women members [17, pp. 311-312].

O'Hara missed the point here. Professionals concerned with the pink-collar effects of the increasing numbers of women among their ranks are rarely preoccupied with whether their salaries are keeping pace with inflation; typically, professionals worry that their status as professionals remains parallel, evidenced by their salaries, with that of other professionals. Thus, although technical communicators' salaries have indeed kept pace with inflation, they have not kept pace, as Stoner demonstrated, with that of comparable professionals such as nurses, engineers, teachers, and social workers.

O'Hara's last statement about the rise in women's membership in STC further clouds the issue. He suggested that we should be pleased that our salaries did not decline when the number of women members rose. Actually, few professions' salaries decline with the feminization of their ranks; they simply do not rise. In other words, technical communication is undergoing exactly the same kind of devaluation that other professions—such as teaching, clerical work, and perhaps soon, medicine—have experienced when the majority of their ranks shifted to women. (For more on feminized occupations and salaries, see Sorensen [16].)

Finally, we should be concerned that women technical communicators' salaries do not equal the salaries of men technical communicators—a point that O'Hara seems to excuse because, after all, statistics show that, nationwide, women earn 70 percent of a man's salary for comparable work, whereas female technical communicators earn 86 percent of their male counterparts' salaries [17, p. 311]. For numerous reasons, we should be concerned about these trends. As technical communicators argue more forcibly for professional status (see, for example, Gordon; Harbaught; Malcolm), their arguments are undermined by these trends toward slower-rising salaries compared with those of other professions [29-31]. Although high salaries do not guarantee professional respectability, low salaries almost always destroy the possibility of increased professional status.

The proportionally lower increase in salaries, of course, is only one of the effects of the feminization of technical communication. As an industry, technical communication may well be affected by the demographics of the roles that women play within corporations. To date, no study has investigated gender-based attitudes and behaviors within the technical communication profession. If there is gender bias within our industry, how does it affect the tasks that technical communicators perform? How does it affect collaborative writing, peer editing, assignment of projects, editorial comments and procedures, information seeking, or presentations? How do male technical communicators react to women as peers, subordinates, supervisors, editors, project managers, engineers, or computer programmers? How do women perceive their male subordinates, supervisors, and peers? Further, we should be aware of other kinds of gender studies that show that women are often tracked into certain kinds of roles, even when they are encouraged to get educations. That women are entering the technical communication profession in increasing numbers may be good news, but how steadily and how quickly are they moving up in the industry? What kinds of occupational roles (researchers, writers, editors, document designers, publications managers, and so on) attract the highest salaries and promotion potential in the technical communication industry? Are women being encouraged to train for these roles and promotions? If so, how? If not, why not?

Some managers will argue that women do not want high-powered, upper-level management jobs because they prefer less stressful roles that allow them to balance their careers with their familial roles. Recent research, however, defies this notion that women want only certain kinds of jobs. In fact, Jacobs found that inclinations "to choose female-dominated occupations . . . are not very stable and change when new opportunities arise" [18, p. 173]. In other words,

> As discrimination is relaxed, women are more than willing to move into a broad range of occupational roles from which they had previously been excluded. This evidence suggests that it is not the lack of interest or 'taste' for work in male dominated occupations that is the cause of women's underrepresentation [18, p. 170].

Further research is needed to investigate questions about the kinds of support women in technical communication find for their aspirations. Is that support different from the kinds of support men find? If it is different, what can the profession as a whole do to encourage fairer treatment of each group? (For research on gender-related trends in promotions, see Solomon; Mandall and Kohler-Gray; DiPrete and Soule; Morrison and Von Glinow; Blum and Smith) [32, 1, 33-35].

Finally, we should compare the role of men and women technical writers who work within our universities and colleges. Although there has long been concern about the hiring, tenuring, and promoting processes regarding all academic technical writers, some studies suggest that particular problems exist for women academicians. In an early study on problems for women in academe, Simon, Clark, and Galway found that although women with PhDs perform as well as men in academe, they may have trouble finding research partners or someone with whom to discuss professional concerns [36]. More significant for technical communication academicians is a 1978 study by Widom and Burke that shows that women see themselves as less likely to write books and articles and more likely to edit these publications [37]. (At my university, and I presume elsewhere as well, writing counts significantly more in a faculty tenure and promotion decision than does editing.) Also significant is that the males in Widom and Burke's study indicated that they felt their publications were above average in comparison to their colleagues' publications, whereas women felt their own publications were below average. These indications of self-image may substantially affect the performance of these professionals, with women feeling a severe disadvantage in the area of their professional lives that is most likely to grant them tenure and promotion. We would do well to replicate this study and others like it to see if any improvements in self-perceptions have occurred with female faculty members and, if so, what effects those changes have had on tenure and promotion decisions. Further, if technical communication as a profession is becoming feminized, what about the academic ranks of technical communication? Is it also increasingly a female-staffed genre? Are our PhD programs graduating more females than males? If so, are they more likely to go into industry or academe? How do their career decisions affect the profession?

Certainly, other questions about the roles of men and women within the technical communication industry exist, and there is much research to be done. A better knowledge of the opportunities and biases of our field will allow stronger recruiting, retention, and productivity from our ranks. And ultimately, recognizing the strengths of all professionals serves only to enhance our professional status.

## TECHNICAL COMMUNICATION RESEARCH

To date, most of our research has focused on broader aspects of communication than on the isolated processes or uses of communication by specific groups of people. This broad approach, however, shortchanges a number of disenfranchised writers, and I suspect astute researchers will find a gold mine of opportunities for research in these narrower areas of process and application. Perhaps a primary reason that gender differences in communication have been ignored for so long is that researchers have had a difficult time separating gender-induced differences from social- or class-based differences. Some of these difficulties are disappearing, making it more likely that studies of gender-based communication are, indeed, measuring what they intend to measure. Still, most of the communication studies focus on speech, not writing those that do focus on writing tend to concentrate on literary writing (see, for example, Sherry; Spender, *Writing*) [38, 39]. The possibilities for more research that addresses functional writing, such as technical communication, certainly hold great promise for scholars in our field. This section presents a sampling of that research, some cautions about the structure and interpretation of research, and some of the possible areas for further research.

Most work on gender and language centers, unsurprisingly, on sexist language because it is so obvious to those of us who are sensitive to language usage. Although researchers have addressed some of the broader issues of nonsexist language usage—primarily stressing how to avoid it (see Vaughn; Wilcoxon; Christian)—we still need to investigate the more particular kinds of awareness and decision processes that most frequently surround the communication of sensitive issues [7, 8, 4]. What kinds of words, implications, and attitudes do people find offensive and why? How can we best educate others who see objections to sexist language as silly? What methods, in other words, can we use to convince people to turn against their habitual biases? Further, how do we measure the presence and extent of sensitivity, tact, or bias in communication? Several works allow us to identify and eradicate sexist language and attitudes, but few attend to these deeper questions about the psychological issues surrounding sexism, sensitivity, and communication.

Since the women's movement helped clarify that women are a group in the sociological sense, sociolinguistic research has documented that women's communication is, indeed, different from men's communication [40, p. 7]. In fact, earlier in this century, women's language was even defined as "abnormal" (quoted in Key) [41, pp. 15-16]. In 1922, Otto Jespersen reported that men had every right to "object that there is a danger of language becoming languid and insipid if we are to content ourselves with women's expressions" [42, p. 246]. Although we now tend to note clearer sociological reasons for the difference between men's and women's communications, the attitude that women's language is corrupt, abnormal, or deviant—an issue I address later in this section—still remains.

Folklinguistics, the study of commonly accepted but undocumented descriptions of language usage, explains many of the stereotypes we have passed down through the centuries about women's and men's speech habits and patterns. Jennifer Coates presented an interesting overview of these stereotypes, focusing on vocabulary, swearing and taboo language, grammatical constructions, literacy, pronunciation, and verbosity [40]. For instance, although most of us are familiar with the stereotype that women are more verbose than men, studies unanimously show that just the opposite is true (see, for example, Eakins and Eakins; Lynch and Strauss-Noll; Swacker; Argyle, Lalljee, and Cook; Wood; Bernard; Chesler) [43-49]. In a discussion of the historical beliefs concerning women's supposed verbosity, Coates said:

> Pre-Chomskyan linguistic enquiry provides us with no evidence that women talked more than men, yet there is no doubt that Western European culture is imbued with the belief that women do talk a lot, and there is evidence that silence is an ideal that has been held up to women for many centuries [40, p. 34].

Coates recounted Dale Spender's interpretation of such impressions: Women who talk at all are judged as talkative, merely because they break the silence, whereas men are judged by the verbosity of other men [40, p. 34].

Much of the current research on sex differences and communication does, in fact, focus on conversation, but much of this research holds useful possibilities for replication as written communication studies. Such research shows, for instance, that women are more likely to ask questions, make statements in a questioning tone ("Your report is due Friday?"), end statements with questions calling for confirmation ("Don't you agree?"), introduce ideas with a question ("You know what we found out?"), and qualify or undercut the strength of their statements ("That's just my opinion") [50, p. 66; 40, pp. 112-113]. Based on the transcripts of 150 court cases, Robin Lakoff presented what is perhaps the seminal study of women's language. She noted several additional qualities of women's language: use of emphatic words ("very," "so," "really"), empty adjectives ("precious," "divine," "charming"), obsessively correct grammar and pronunciation, direct quotations, and use of special vocabulary—especially for colors [51]. Although these observations are clearly based on conversational tendencies, we need to investigate whether they have correlations in written communications. Particularly valuable should be research on women's written descriptions of mechanisms and claims for product performance with regard to their use of adjectives, hyperbole, substantiating support for claims, and qualifying phrases.

Researchers have also shown that in conversation men are five times more likely to interrupt women than women are to interrupt men (see, for example, Zimmerman and West; West and Garcia; Frost; Segal and Tower) [52-55]. Interestingly, the reversal of such conversation patterns—women interrupting men—is interpreted negatively by men. Joyce Hocker Frost found that when typical conversation patterns are reversed so that women interrupt men, men perceive the women as attacking them or, more interestingly, as being in a bad mood [54, p. 132]. When women are the victims of abrupt changes in topics or interruptions by men, however, the women tend to become silent rather than protesting or ignoring the attempted change [40, p. 117]. Because of these differences in men's and women's communication styles, further research needs to address questions about how these qualities affect women in the industrial or academic applications of technical communication. How do they affect women's confidence, their roles, and their permitted contributions? For example, if women are typically interrupted when presenting their ideas about, say, organizing or formatting a document, then what is the result of that interruption? Do female technical communicators, in effect, learn to keep their ideas to themselves and is that the behavior their male counterparts want to perpetuate? Should a woman persevere and insist on being heard in a collaborative setting? What is likely to be the perception of such a tactic—that the woman is goal-oriented and a good contributor or that she is pushy and counterproductive?

Other aspects of women's language are also garnering attention from sociolinguists, psychologists, and communication researchers. Most readers are probably not surprised to find that women use more stroking language—encouraging everyone to feel good about a project, plan or role (see Lynch and Strauss-Noll)—whereas men use more inspirational/competitive language to encourage everyone to be ready for the next battle with the competition [44]. Such predispositions also affect the ways that men and women handle conflict. As Frost reported:

> I find that women use more accommodative conflict strategies, while men use competitive or exploitive ones. Men are more comfortable with winning, while women are more comfortable with finding a fair outcome, acceptable to all. Neither of these strategies is inherently 'good.' Sometimes the structure of a conflict is set up for winning and losing, and sometimes the structure is cooperative. The problem occurs when men and women get stuck in a role and cannot change when it is appropriate to change [54, p. 133].

Because women are more likely to encourage harmony, they are also less likely to use confrontational language; thus their language is typically characterized as more apologetic and as free from threats, ultimatums, or even biased implications or statements. Given that certain tasks in technical communication—editing, for instance—require practitioners to be forceful, confident, and even confrontational at times, we should wonder if women's struggles for harmony affect the jobs that women editors do. On the other hand, women's tendency to promote harmony may lead them to be better negotiators, allowing them to see both sides of an issue rather than feeling compelled to win each editorial bout. Thus we might want to investigate how successful women editors approach the editorial work they do, the kinds of comments and notations that women make versus those made by their male counterparts, and the steadfastness of women versus men writers who are challenged by editors. Are women more likely to make changes just to promote harmony?

Clearly, we need to investigate whether any of these aspects of oral communication are also aspects of written communication. Unfortunately, we have yet to create the kinds of instruments that will appropriately measure some of the more subtle nuances of language. Although we typically know a compliment or threat when we hear one, we have yet to find some reliable way to measure objectively the inclusion or effects of word choice, sentence length, and even punctuation on various kinds of communication. Such a measure might allow us to determine different subtleties of communications, ranging from a warning—"I wouldn't do that if I were you"—to a threat—"If you do that, you'll be fired." Further, such a measure might allow us to assess the gender implications of statements while also pointing out successful communication strategies. And those benefits may well lead us to an understanding of the characteristics of androgynous communication.

Elizabeth Tebeaux made similar observations in her recent study on the sensitivity and success in communicating unpopular messages, where she also noted that the effects of gender differences on writing vary according to how much "people-intensive work experience" the writers have had [56, p. 27]. Thus writers with more experience dealing with people frequently wrote more sensitive and successful communications; their experience seems to have suppressed gender distinctions. Based on Tebeaux's initial observations, writers' abilities to write sensitive and successful communications may also be positively correlated with maturity—not just age, but attitudes, adaptability, responsibility, and level of tolerance as well. Such distinctions of maturity may help us structure academic programs that address communicative sensitivity at appropriate times in a student's development.

Thus researchers need to investigate the particular qualities of women's language that distinguish it from men's. We also need to distinguish the kinds of research and assumptions that still apply to our understanding of communication differences from those that stereotype communicators and rhetorical situations. In a 1975 work, for instance, Key showed that women's language rarely includes explanations when speaking to men—rather, men explain to women. As she pointed out, "the male who uses explanation is not really interested in the female acquiring more knowledge .... Rather, he is showing his superiority" [41, p. 37]. Surely, technical communication is attempting to change these assumptions, so a study of strategies and purposes for communicating with members of same sex versus members of the opposite sex would make an especially useful study in technical communication.

In addition to the issues of language usage, we should also investigate whether the writing processes of women technical communicators vary from those of their male counterparts. Do women, for instance, spend more time in the research phase, the prewriting phase, the writing phase, or the revising phase than do their male counterparts? Do they undertake these phases with different goals in mind? Do they make editorial changes based on the same kinds of reasons, preferences, or instincts as men? Do they typically make more or less changes than do their male counterparts? Do they evaluate their work differently?

Of the works that have found a place in our literature, some unfortunately introduce or perpetuate unclear or stereotypical thinking about gender and gender-related communication differences. Smeltzer and Werbel addressed some of this unclear thinking and its related research by pointing to Barryman's remark: "Although very few actual differences in the communication of males and females are empirically documented, stereotypical assumptions, perceptions, and expectations concerning the linguistic behavior of the sexes persist" (quoted in Smeltzer and Werbel) [57, p. 42]. In fact, in a limited study, Smeltzer and Werbel's research "refutes beliefs that any qualitative difference may exist between the communication of men and women" [57, p. 47]. More substantial studies may confirm their suspicions, but we should not discount the possibilities and implications of differences until more pervasive research is undertaken.

Some problems with ambiguity are indeed creeping into what little work has been done with gender studies and technical communication. Nancy Veiga presented a fine overview of the gender communication issue but generates some illogical conclusions [58]. For instance, the 1978 research by Eakins and Eakins that Veiga used to support her claims shows that men take more turns speaking and use more words than do women. Based on these findings, Eakins and Eakins concluded that "males seem to be more sensitive to success or failure of communication than females" (quoted in Veiga) [58, p. 279]. Surely those of us in technical communication would question this conclusion: Perhaps, women are simply capable of making their points more concisely and directly than their male counterparts, or perhaps they are interrupted and never get the chance to complete their turn at speaking. In either case, the relationship between number of words used and number of turns speaking does not necessarily correlate with sensitivity to success or failure of communication, although, of course, whether women are heard when they speak or allowed to finish speaking is directly related to the success or failure of communication. Other conclusions in Veiga's work are also questionable. She stated:

> It is ironic that society perceives women as more verbose than men, given the Eakins' analysis, which proves, at least in their experience, that men talk longer and more often than women. As women ascend the corporate ladder, they must develop superb written and oral skills to overcome sex stereotyping [58, pp. 281-282].

Although each of these sentences may be true, their juxtaposition implies several bizarre conclusions: (a) Women must be more like men to ascend the corporate ladder; (b) talking longer and more often is desirable; (c) talking longer and more often is evidence of superb oral skills; and (d) talking longer and more often will help women overcome sex stereotyping.

In fact, much of the research on women's and men's communication strategies and practices has been riddled with stereotypes leading to erroneous conclusions and abandoned testing. Spender explained:

> One indictment of this research area is that so many of the hypothesized differences that have been tested have not been found. This is not necessarily because research techniques are unsophisticated and inadequate and therefore incapable of locating sex differences in language use: it is primarily because research procedures have been so embedded with sexist assumptions that investigators have been blinded to empirical reality. Sexist stereotypes of female and male talk have permeated research and often precluded the possibility of open-ended studies which may have revealed sex differences—and similarities—in language usage *(Man Made Language)* [59, pp. 32-33].

For example, researchers and pop psychologists have presumed that women use more tag questions ("Don't you agree?") than do men because women are insecure. In some research, however, men have been shown to use more tag questions than do women (Dubois and Crouch) [60]. Yet, as Spender pointed out, the pattern was never cited as a suggestion "that it is men who might lack confidence in their language" [59, p. 9]. According to Spender, the significance of this bias is that researchers still assume that women's language is deviant and men's language is normal [59, p. 8]. She continued by noting that researchers "focus on small segments of female speech and the conviction that if investigators look long enough and hard enough—and in the 'right' place—they are bound to find these hypothesized deficiencies in female speech" [59, p. 33]. Clearly, we need to be more careful in the ways we construct research on gender communication and in the ways we interpret findings by returning to the basics of research methods that stress validity and reliability (see Benderly; Baumeister; Fagley and Miller; Rothblum) [61-64].

We might also try to explain the contradictions inherent in some unrelated studies. For instance, White's study shows that throughout their educational experiences, females write better than men; an unrelated study by Bowman shows that communication skills are ranked as the greatest asset for getting promotions and rising to the top of the corporate ladder; and still other studies show the paucity of women at the top of the corporate ladder—"only 2% of the top executives in Fortune 500 companies are women" [15, 65; 32, p. Solomon 96]. When we put the conclusions of these studies together, we end up with an interesting contradiction: If women are better communicators than men, and communication skill is the primary asset for management positions, then why are so few women in managerial positions? Clearly, there must be some other reason for women's failing to break into the ranks of top-level management. Thus, beyond broadening the extent of the questions we ask and trying to get beyond surface observations and solutions or quick-fix remedies, we also need to develop clearer interpretive skills of the data our research generates and compare the ideas accumulated from different kinds of research.

## TECHNICAL COMMUNICATION PEDAGOGY

All of this research should lead to more informed teaching of technical communication principles—both theory and practice. We may need to learn different techniques for presenting ideas to different groups of readers, but primarily we should be concerned that we do not perpetuate gendercentric views of communication. The diversity we should promote takes two forms: in the ways we discuss technical communication with our students and in the nature of the assignments we give them.

First, we need to acknowledge the ways that gender issues may affect what we teach. For instance, we need to reevaluate our teaching of audience analysis to focus on different groups of readers and their needs beyond notations of their education and job levels. Rather than limiting audience analysis to these simplistic issues, we need to encourage our students to see audience analysis as an extremely complex matter that may have to accommodate gender and its effects on the readers' backgrounds, experiences, and expectations. Joseph Ceccio and Michael Rossi undertook such a view in their research on audience analysis. They acknowledged that "good audience analysis involves a conscious, planned, goal-directed use of stereotypes; its worst enemy is the unconscious, untested, preconceived notion or bias" [66, p. 78]. As Ceccio and Rossi suggested, audience analysis can be enhanced through the writer's relevant perceptions of gender roles. Without determining the importance (or unimportance, perhaps) of the audience's gender roles in interpreting or understanding any particular piece of information, writers are essentially paralyzed. For example, if men are the primary audience for a report or presentation on infant care, what kinds of assumptions can be made about their knowledge and needs? We may, perhaps, assume that many of them have been socialized to have little contact with infants and may feel uncomfortable around babies; consequently, our first activity may be getting each man to hold an infant. This example shows the kinds of constructive decisions we can make about audiences and their needs for information. Further, this analysis does not insist that no man has ever held an infant; thus nothing in the assumption or the resulting approach to a lesson on infant care should cause miscommunication by being offensive.

We also need to incorporate gender considerations into other areas of our teaching, especially in the assignments we give our students. A review of the suggested topics for writing assignments presented in our technical communication textbooks suggests that students have rarely been encouraged to undertake projects that require more than the basic reporting of facts. For instance, these textbooks typically suggest that students write reports that compare computers and recommend the best one for the XYZ Corporation, analyze the effects of computerizing the company's inventory system, or investigate whether a company should offer a new service or product. Although it is entirely plausible for an employee to be asked to research and write these kinds of reports—in fact, they may well be the most frequently requested reports—students should not need an entire course to teach them to handle such requests. I suspect, in fact, that students could manage such a project at the end of a composition course—an assignment that would give composition teachers a chance to prove the relevance of their course, while giving students an opportunity to apply what they have learned to "real world" assignments. Thus one of the first areas of investigation and accommodation we in technical communication need to undertake is to reevaluate the kinds of assignments we give our students. An essential task may well be to provide particular kinds of writing exercises or projects that reflect the effects of gender issues in technical communication or in corporate settings. In addition, rather than reporting their basic investigations, students in technical communication courses should be studying the more complex aspects of communication that require them to develop and demonstrate resourcefulness, openmindedness, and tact.

In particular, students would benefit from assignments that challenge or develop their attitudes and that require them to respond to complex communications situations. Reports on gender relations within a company or a particular department require such clearheadedness. So do particular issues that affect gender relations: day-care accommodations, maternal and paternal leave, flextime, and other alternative workstyle arrangements for women and men who do not fit the traditional nine-to-five workday (see Murphree) [67]. Increasingly, American corporate executives are recognizing the harm they do to their corporations when they refuse the services of intelligent, capable, hardworking people merely because they cannot accommodate a traditional schedule. Gender issues certainly affect the contemporary workplace, but technical writing assignments have not reflected these changes. Investigations of alternative workstyles, dress codes, sexual harassment policies, employee benefits, hiring and promotion policies, and so on present excellent opportunities for students to research and report inclusive, rather than exclusive, issues that affect business and industry. Proposals for developing flexible work schedules, day-care accommodations, or sensitivity workshops may lead to reports that highlight the results of these investigations. Other reports may be generated from studies of feasibility, implementation, or even attitudes about such services or changes in the workplace.

Other writing assignments are also possible. Letter or memorandum assignments, for instance, could require students to create or address,

- inquiries about sexual harassment policies
- complaints of sexual harassment
- complaints of unfair hiring/promotion practices
- inappropriate behavior toward any worker
- announcements of sensitivity workshops
- ideas for the establishment of gender-neutral language in publications
- ideas for the establishment of gender-neutral social activities.

And, of course, excellent writing assignments can be structured as responses to any of these letters or memos. Students who intend to pursue business careers rarely get an opportunity even to discuss these kinds of issues, much less respond to them in writing. Assignments that require responses to these issues allow students to demonstrate more than reporting and formatting knowledge; they provide opportunities for the students to demonstrate the kinds of management and leadership decisions and communications the students would actually execute within an organization. In other words, such assignments allow students to broaden their portfolios, while showing prospective employers the kind of management style that they have adopted and the communication-based manifestations of that style.

## CONCLUSION

Throughout this article, I have tried to demonstrate areas of research that are available to us through gender concerns. We need to investigate our profession—both the technical communication industry and academe—research, and pedagogy to find ways to accommodate the changing corporate culture of American businesses. As professional technical communicators in industry and academe, we should be concerned that we may be overlooking or even misusing someone's potential, rewarding characteristics we are comfortable with rather than what is most productive, sensible, or creative. We should be concerned in our quest for professional status that we are, indeed, professionals and that we consistently acknowledge and promote those who are most valuable to our efforts as communicators. In our research, we should study the processes and practices of various writers, withstanding the inclination to assign value to any particular method except as it reflects productivity and quality. In our classrooms, we must encourage excellence, regardless of gender. Further, we must promote students' understanding and tolerance of the workplace's diversity as well as their attention to the propriety of language and attitudes within the workplace and the kinds of gender issues that affect workers. Finally, we must see to it that these issues become thoroughly integrated into our curriculum, rather than composing a special unit of instruction that is taken up for a week or two and then dropped.

## REFERENCES

1. B. Mandall and S. Kohler-Gray, Management Development That Values Diversity: Unbiased Selection of Management Candidates, *Personnel*, pp. 41-47, 1990.
3. M. M. Lay, Interpersonal Conflict in Collaborative Writing: What We Can Learn from Gender Studies, *Journal of Business and Technical Communication, 3*:2, pp. 5-28, 1989.
4. B. Christian, Doing Without the Generic He/Man in Technical Communication, *Journal of Technical Writing and Communication, 16*, pp. 87-98, 1986.
5. W. Martyna, Beyond the He/Man Approach: The Case for Nonsexist Language, in *Langauge, Gender and Society*, B. Thorne, C. Kramarae, and N. Henley (eds.), Newbury, Rowley, Mississippi, 1983.
6. M. Z. Corbett, Clearing the Air: Some Thoughts on Gender-Neutral Writing, *IEEE Transactions on Professional Communication, 33*:1, pp. 2-6, 1990.
7. J. Vaughn, Sexist Language—Still Flourishing, *The Technical Writing Teacher, 16*, pp. 33-40, 1989.
8. S. A. Wilcoxon, He/She/They/It?: Implied Sexism in Speech and Print, *Journal of Counseling and Development, 68*, pp. 114-116, 1989.
9. R. B. Stoner, Economic Consequences of Feminizing Technical Communication, *Proceedings of the 35th International Technical Communication Conference*, Society for Technical Communication, Washington, D.C., 1988.
10. J. A. Daly, R. A. Bell, and J. Korinek, Interrelationships among Attitudes toward Academic Subjects, *Contemporary Educational Psychology, 12*, pp. 147-155, 1987.
11. C. Holden, Female Math Anxiety on the Wane, *Science, 236*, pp. 660-661, 1987.
12. S. P. Marshall and J. D. Smith, Sex Differences in Learning Mathematics: A Longitudinal Study with Item and Error Analysis, *Journal of Educational Psychology, 79*, pp. 372-383, 1987.
13. D. Berliner, Math Teaching May Favor Boys over Girls, *Education Digest, 53*:3, p. 29, 1988.
14. S. Clement, The Self-Efficacy Expectations and Occupational Preferences of Females and Males, *Journal of Occupational Psychology, 60*, pp. 257-265, 1987.
15. J. White, The Writing on the Wall: Beginning or Ending of a Girl's Career?, *Women's Studies International Forum, 9*, pp. 561-574, 1986.
16. E. Sorensen, Measuring the Pay Disparity between Typically Female Occupations and Other Jobs: A Bivariate Selectivity Approach, *Industrial and Labor Relations Review, 42*, pp. 624-639, 1989.
17. F. M. O'Hara, Jr., Trends in STC, *Technical Communication, 36*, pp. 310-312, 1989.
18. J. A. Jacobs, *Revolving Doors: Sex Segregation and Women's Careers*, Stanford University Press, Stanford, California, 1989.

19. B. Sarkis, The Competency Factor: Results are What's Important, *Credit and Financial Management*, p. 12, September 1987.
20. J. E. Coverdill, The Dual Economy and Sex Differences in Earnings, *Social Forces, 66*, pp. 970-993, 1988.
21. V. Bozzi, The Low Track Wage, *Psychology Today*, pp. 98-110, February 1988.
22. J. D. Olion, D. P. Schwab, and T. Haberfeld, The Impact of Applicant Gender Compared to Qualifications on Hiring Recommendations, *Organizational Behavior & Human Decision Processes, 41*, pp. 180-195, 1988.
23. B. Gerhart, Gender Differences in Current and Starting Salaries: The Role of Performance, College Major, and Job Title, *Industrial and Labor Relations Review, 43*, pp. 418-433, 1990.
24. M. Tienda, S. A. Smith, and V. Ortiz, Industrial Restructuring, Gender Segregation, and Sex Differences in Earnings, *American Sociological Review, 52*, pp. 195-210, 1987.
25. C. C. Fischer, Toward a More Complete Understanding of Occupational Sex Discrimination, *Journal of Economic Issues, 21*:2, pp. 113-138, 1987.
26. P. Glick, C. Zion, and C. Nelson, What Mediates Sex Discrimination in Hiring Decisions? *Journal of Personality and Social Psychology, 55*, pp. 178-186, 1988.
27. C. Goldin and S. Polachek, Residual Differences by Sex: Perspectives on the Gender Gap in Earnings, *American Economic Review, 77*:2, pp. 143-151, 1987.
28. J. Muniz, E. Garcia-Cueto, E. Garcia-Alcaniz, and M. Yela, Analisis de la Fluidex Verbal, Oral y Escrita en Hombres y Mujeres, (Analysis of Verbal, Oral, and Written Fluency in Men and Women), *Revista de Psiologia General y Aplicada, 40*, pp. 255-275, 1985.
29. K. M. Gordon, Do We Deserve Professional Status? *Technical Communication, 35*, pp. 268-274, 1988.
30. F. W. Harbaught, Professional Certification: To Be or Not To Be, *Technical Communication, 36*, pp. 93-96, 1989.
31. A. Malcolm, On Certifying Technical Communicators, *Technical Communication, 34*, pp. 94-102, 1987.
32. C. M. Solomon, Careers Under Glass: Barriers to Advancement, *Personnel Journal*, pp. 96-105, April 1990.
33. T. A. DiPrete and W. T. Soule, Gender and Promotion in Segmented Job Ladder Systems, *American Sociological Review, 53*, pp. 26-40, 1988.
34. A. M. Morrison and M. A. Von Glinow, Women and Minorities in Management, *American Psychologist, 45*, pp. 200-208, 1990.
35. L. Blum and V. Smith, Women's Mobility in the Corportation: A Critique of the Politics of Optimism, *Signs, 13*, pp. 528-545, 1988.
36. R. J. Simons, S. M. Clark, and K. Galway, The Woman Ph.D.: A Recent Profile, *Social Problems, 15*, pp. 221-236, 1967.
37. C. S. Widom and B. W. Burke, Performance, Attitudes, and Professional Socialization of Women in Academia, *Sex Roles, 4*, pp. 549-562, 1978.
38. R. Sherry, *Studying Women Writing: An Introduction*, Arnold, New York, 1988.
39. D. Spender, *The Writing or the Sex? Or Why You Don't Have to Read Women's Writing to Know It's No Good*, Pergamon, New York, 1989.

40. J. Coates, *Women, Men, and Language: A Sociolinguistic Account of Sex Differences in Language*, Longman, New York, 1986.
41. M. R. Key, *Male/Female Language: With a Comprehensive Bibliography*, Scarecrow, Metuchen, New Jersey, 1975.
42. O. Jespersen, *Language: Its Nature, Development, and Origins*, Allen & Unwin, Winchester, Massachusetts, 1922.
43. B. W. Eakins and R. G. Eakins, *Sex Differences in Human Communication*, Houghton Mifflin, Boston, Masscahusetts, 1978.
44. C. M. Lynch and M. Strauss-Noll, Mauve Washers: Sex Differences in Freshman Writing, *English Journal, 76*:1, pp. 90-94, 1987.
45. M. Swacker, The Sex of the Speaker as a Sociolinguistic Variable, in *Language and Sex: Difference and Dominance*, B. Thorne and N. Henley (eds.), Newbury, Rowley, Mississippi, 1975.
46. M. Argyle, M. Laljee, and M. Cook, The Effects of Visibility on Interaction in a Dyad, *Human Relations, 21*, pp. 3-7, 1968.
47. M. Wood, The Influence of Sex and Knowledge of Communication Effectiveness on Spontaneous Speech, *Word, 22*:1, 2, 3, pp. 112-137, 1966.
48. J. Bernard, *The Sex Game*, Atheneum, New York, 1972.
49. P. Chesler, Marriage and Psychotherapy, *The Radical Therapist*, The Radical Therapist Collective (eds.), Ballantyne, New York, 1981.
50. A. Kohn, Girl Talk, Guy Talk, *Psychology Today*, pp. 65-66, February 1988.
51. R. Lakoff, *Language and Woman's Place*, Harper & Row, New York, 1975.
52. D. H. Zimmerman and C. West, Sex Roles, Interruptions, and Silence in Conversation, *Language and Sex: Difference and Domination*, Newbury, Rowley, Mississipi, 1975.
53. C. West and A. Garcia, Conversational Shift Work: A Study of Topical Transitions between Men and Women, *Social Problems, 35*, pp. 551-575, 1988.
54. J. H. Frost, The Influence of Female and Male Communication Styles on Conflict Strategies: Problem Areas, *Women, Communication, and Careers*, M. Grewe-Partsch and G. J. Robinson (eds.), Saur, New York, 1980.
55. J. A. Segal and R. S. Tower, Substance versus Style, *Personnel Administrator*, pp. 101-104, August 1989.
56. E. Tebeaux, Toward an Understanding of Gender Differences in Written Business Communication, *Journal of Business and Technical Communication, 4*:1, pp. 25-43, 1990.
57. L. R. Smeltzer and J. D. Werbel, Gender Differences in Managerial Communication: Fact or Folk-Linguistics? *Journal of Business Communication, 23*:2, pp. 41-50, 1986.
58. N. E. Veiga, Commentary: Sexism, Sex Stereotyping, and the Technical Writer, *Journal of Technical Writing and Communication, 19*, pp. 277-283, 1989.
59. D. Spender, *Man Made Language*, (2nd Edition), Routledge and Kegan Paul, Boston, Massachusetts, 1985.
60. B. L. Dubois and I. Crouch, The Question of Tag Questions in Women's Speech: They Don't Really Use More of Them, Do They? *Language in Society, 4*, pp. 289-294, 1975.

61. B. L. Benderly, Don't Believe Everything You Read . . . A Case Study of How the Politics of Sex-Difference Research Turned a Small Finding into a Major Media Flap, *Psychology Today*, pp. 67-69, November 1989.
62. R. F. Baumeister, Should We Stop Studying Sex Differences Altogether? *American Psychologist, 43*, pp. 1092-1095, 1988.
63. N. S. Fagley and P. M. Miller, Investigating and Reporting Sex Differences: Methodological Importance, the Forgotten Consideration, *American Psychologist, 45*, pp. 297-298, 1990.
64. E. D. Rothblum, More on Reporting Sex Differences, *American Psychologist, 43*, p. 1095, 1988.
65. G. W. Bowman, What Helps or Harms Promotability? *Harvard Business Review*, p. 6+, January-February 1964.
66. J. F. Ceccio and M. J. Rossi, Inventory of Students' Sex Role Biases: Implictions for Audience Analysis, *The Technical Writing Teacher, 8*, pp. 78-82, 1981.
67. C. T. Murphree, How Far Have You Come, Baby? *Bulletin of the Association for Business Communication, 51*:1, pp. 29-31, 1988.

# CHAPTER 5

# Ethics

Ethics until recently has been a relatively neglected area of technical communication. In the last two decades, however, the need for the ethical use of technology and science has become glaringly apparent. We are all familiar with dramatic lapses in ethical conduct in technical areas such as Bhopal, Three Mile Island, and Challenger. Beyond such specific incidents, however, there also has arisen a pervasive general concern for future generations and for the environment, for the ethical stewardship of the Earth itself. With our awareness of such lapses and the grave threats to our habitat have come calls for the increased integration of ethics in technical communication scholarship, pedagogy, and practice.[1]

Reflecting such concerns, practically every technical communication textbook now contains a section on ethics. In addition, scholarly and professional journals such as *IEEE Transactions on Professional Communication* (*30*:3, 1987) and *Technical Communication* (*27*:3, 1980) have published special issues on ethics. The Society for Technical Communication, in addition, has published an entire book on this subject (reviewed later).

The connection between ethics and technical communication, though, has been problematic. From before the time of Aristotle, some have understood ethics to be separate from technology and science. That is, definite, specialized knowledge such as from technology and science has been seen as inherently separable from such indefinite, controvertible activities as rhetoric and ethics. In more recent years, however, critics have increasingly recognized that the dualistic separation of science and technology from ethics is both unwarranted and harmful. As a result,

---

[1] For a general reference book on ethics from business, professional, and medical perspectives that is highly pragmatic, situational, and social constructionist, I recommend *Medical Ethics: A Reader* by A. Zucker et al. [1]. For a general reference book on ethics from a philosophical perspective, I recommend P. Singer (ed.), *A Companion to Ethics* [2].

the current curricula of many technical and scientific institutions such as Rensselaer Polytechnic Institute, Carnegie Mellon Institute of Technology, and the Massachusetts Institute of Technology have been shaped to mend this partition.

In this chapter I will first discuss the definition of ethics and how it is particularly problematic for technical communication. Next I will review the literature on this topic and offer theoretical and critical commentary. Last I will introduce the reprinted essays.

## PROBLEMS OF DEFINITION AND USAGE

Writers typically begin their discussions of a topic by offering at least a provisional definition that is then consistently carried throughout the work. Thus we as readers know precisely what the writer is writing about. With ethics, however, definition is problematic. In this section, therefore, I will not argue for a single, particular, consistent definition of ethics because to confine myself to a single definition would impede consideration of alternate perspectives and definitions.

"Ethics" has many referents. It can refer to the general study of goodness or rightness. It can also refer to a particular system or theory of goodness and rightness, either explicitly defined or taken for granted. Another common usage is as a principle that tacitly informs an activity, such as frequently said of capitalism, Marxism, or particular technologies. Thus "ethic" is used regarding technology and science in general, such as in J. Ellul's critique of technologism and R. Weaver's critique of scientism (not the religious denomination but science as a world view) [3, 4].

Others, however, oppose such usage, arguing that technology and science are specifically separate from arguments about goodness and rightness; though one can say that particular *values* such as efficiency, durability, and feasibility inhere in science and technology, there is no systematized consideration of goodness and rightness in them and so "ethics" should not be used regarding them. Still others object precisely to this line of argument, holding that it without justification elevates technology and science above ethics, and in so doing gives them free reign to be used in any and all directions. This, for example, is the objection of S. Monsma and J. Ellul [5, 3].

Because of these persistent and irresolvable difficulties, the reader is asked to tolerate indefiniteness and ambiguity in this discussion for the sake of comprehending the great variety of opinions on this topic. It is a fundamental (though understandable) mistake to assume that ethics can be reduced to a limited set of definitive points which can then be systematically applied, that is, that it can be treated technically.

In the following sections, I will explore the questions of the ethics of technology and science in themselves; the relation of ethics to rhetorical theory; the idea of the ethics of nature; conformance to good practice; and codes of conduct.

## Do Science and Technology have Intrinsic Ethics?

Perhaps the most fundamental question is, Do they have intrinsic ethics? Many say, Yes. The continuing revelation of the social constructedness and the rhetoricity of science and technology show them to be social enterprises moved by persuasion on the basis of certain values. In addition, other scholars are revealing numerous instances of gender-bias in technology and science, either in principle or in practice. Thus many scholars now accept that science and technology seem to entail values of their own and so it is appropriate to refer to them as embodying an ethic.

There are many practical ramifications of viewing science and technology as ethical systems. A. G. Gross, the rhetorician, for example, explains that science does not have the emotional disengagement it often claims for itself [6]. Some feminists, too, think that the basic principle of objectivity in science entails treating humans as objects, which is potentially damaging to people.[2] Feminism, indeed, itself presents a problem for ethics. Some feminist theorists argue that there is a particularly feminist ethic distinct from traditional ethics, which they identify as masculinist. Gilligan emphasizes sociality, compassion, and contingency in contrast to the male-characteristic values of individuality, justice, and absoluteness [8]. Similarly, Noddings emphasizes caring in her feminine approach to ethics and moral education [9].[3]

Considering ethics classically also can be illuminating. Classically, ethics involved the pursuit of the good, true, and right, good being what is sought for its own sake, true being what is absolute and unchanging, and right being what is just and pleasing to the gods. Though this is hardly the language of technical communication textbooks, these notions are still active implicitly in almost every discussion of technical communication. The good and right appear in criteria of evaluation as well as in codes of conduct and standards of sound practice. The true appears, too, at least implicitly, in the form of the absoluteness and immutability of mathematical formulae and physical laws.[4]

---

[2] For an in-depth discussion of the distinction between objectivity and objectification (what others call objectivization), see D. Dobrin's *Is Technical Writing Particlarly Objective?[7]*.

[3] A good source on the on-going debate as to whether there is or should be an ethic specific to either feminists or women is the collection of essays edited by M. J. Larrabbe, *An Ethic of Care* [10].

[4] For a quasi-scientific approach to ethics, see T. Trainer who considers morality as socially defined and constructed [16]. This makes morality amenable to a scientific investigation of sorts. The reader should see, however, the implicit elevation of science above conventional morality as well as the reductivist thrust of this approach.

## Ethics and Rhetorical Theory

Contemporary rhetorical theory holds that language, and all knowledge constituted and mediated by language, always inescapably embodies, represents, and propagates a world view and therefore a system of values. Thus science and technology embody ethics by virtue of their rhetoricity.

K. Burke, for example, explains that we view the world through "terministic screens" represented in our language and cannot do otherwise; we can only exchange one screen for another [11]. R. Weaver explored the inescapable ethicality of language and pointed out that many ideas such as "science" and "progress" are taken in practice as so presumably true and absolutely desirable that they carry a sort of religious rightness to them—"god terms," he called them [4]. S. M. Halloran argues for a civic yet personalistic rhetoric which emphasizes speaking and writing in a civically effective and responsible way as an intellectual virtue integral to and inseparable from the particular person, making communication ethically responsible. P. Bizzell explains that the stance of neutrality, objectivity, and scrupulous relativism often sought by communication professors is an actual impossibility and, paradoxically, represents an ethic itself [13].

For Burke, Bizzell, Weaver and Halloran, then, a world view entails an ethic, even when it is the mistakenly supposed view that science and technology are removed from ethics. If all language and rhetoric is inherently ethical, and all science and technology is mediated by rhetoric, then science and technology unavoidably involve ethics.

In technical communication, one of the expression of these rhetorical ideas is the rejection of the "transparency" view of language.[5] Miller and Halloran both, for instance, trace the history of the erroneous transparency view through the establishment of the Royal Academy of Sciences, and the pervasiveness of this assumption among contemporary scientists, technicians, and technical communicators [15, 11].

Among the effects of holding this erroneous assumption, they point out, is a view of language and communication that is highly suspicious and disparaging. They explain that if one insists that communication is best when it is transparent and unobtrusive, the implicit assumption is that language only gets in the way of the real purpose of communication, the relaying of information and the presentation of facts. As discussed earlier, this assumption was operative in the founding of the Royal Academy of Science, which viewed language only as empty, if not deceptive, embellishment. This assumption was vigorously pursued again later by

---

[5] For a clear description and critique of the "transparency" view of language, see C. R. Miller's *A Humanistic Rationale for Technical Writing* [15].

the positivists of the late nineteenth and early twentieth centuries who tried to avoid language altogether in communication by working only with facts, symbols for facts, and the mathematical and logic interrelations among them. This movement to supercede word-based language with the pure symbols of mathematics and logic failed because it assumed language was the source only of confusion; it failed to recognize that language is itself an important source of knowledge in constituting our conceptions and representations. As Burke pointed out repeatedly, our "terministic screens" not only shape the world we perceive but also shape our own identities.

## Ethics from Nature and the Example of Frank Lloyd Wright

The lawfulness of nature is sometimes understood as a sort of ethic, as technical rules cannot be violated.

The life of Frank Lloyd Wright, the famous American architect, illustrates this lawful authority of nature [17]. One of Wright's chief developmental experiences as an architect occurred when as a boy he wanted to built a windmill, which he built it after his own intuition. It failed the test of Nature, however, when a severe storm reduced it to rubble. This experience taught Wright the need to conform to the laws of Nature regardless of one's intuitions or enthusiasms. He learned, in effect, the ethics of Nature.

Another instance in Wright's life taught ethics in the sense of rules of conduct defined by humans. While Wright was living in Madison, Wisconsin, the state capitol building collapsed during construction, killing several people. The contractor had used inferior materials in the foundation, against prevailing engineering practice. This taught Wright that engineers had developed another set of rules which also must be learned and conformed to. Thus Wright resolved to master engineering before turning to architecture. Though ethics entails at the very least conformance to the laws of both Nature and principles of good practice, many critics feel that conformance alone, discussed next, is not completely adequate.

## Conformance to Good Practice

The usual understanding of ethicality as conformance is fairly straightforward: one should do only "good" science and technology, the doing of which in itself satisfies the need for ethicality. Following sound engineering practices, thus, is taken as an expression of responsibility and ethical concern.

To see such conformance to good practice at work, consider the Challenger disaster. Two of the findings of the Rogers Presidential commission that investigated the disaster were that the design of the joint between segments of the solid rocket booster was basically inappropriate for its function and deviated from standard engineering practice, and that the material of the O-ring seals likewise was inappropriate for the conditions under which would operate, another deviation from good practice [18]. Conforming to standard procedures and good practice, then, involves an ethic in itself.

## Codes of Conduct

Though codes of conduct have a comfortable definiteness about them and certainly have an important function to play, they cannot of themselves cover some important areas of ethical concern. Ethics is obviously not an object because it is an abstract, subjective construct, but neither can it be objectified fully into a list of specific, absolute rules. This is not to say that codes of ethical conduct are useless, for they indeed function at least as general outlines or categorizations of ethical expectations. They are of themselves, however, never sufficient and never a substitute for the individual or corporate weighing of the consequences of one's actions.[6]

One of the problems with codes or systems of procedures is that they are reductive: they assume that complex practices and particular circumstances can be reduced effectively to a finite set of principles or procedures. This assumption simply denies the complexity and contingency of real life and the uniqueness of particular instances. Another major problem is that they assume a sort of prescience such that one can anticipate with assurance the state of affairs of the future and all the forces and contingencies operative then. The questionable assumption of prescience is, for example, one of the criticisms leveled against the software developed for the Strategic Defense Initiative.

A related difficulty also involves the complexity of real events but in a different way. These complexities can be "covered," it is thought, by making codes or systems of procedures very general. Such codes, however, then suffer from the difficulty of being so vague and indefinite as to be applicable only with great difficulty to particular instances. They can also end up sounding almost platitudinous and beg the question of what is good or right to do in day-to-day activities.[7]

Furthermore, conformance as a standard is not itself wholly satisfying because it equates ethics with conformance. Reference to codes or systems of procedures, however, can obscure or even deliberately obfuscate ethicality. By pointing to an

---

[6] For a discussion of codes of conduct that reviews many sources and is more critical than my own exploratory discussion, see Markel [19].

[7] For further discussions of altruism and vagueness in codes, see Buchholz [20].

external referent, a fixed code, one moves the burden of responsibility from oneself to an impersonal entity. S. Katz, for instance, indicates how conformance to the external referent of expediency (or utility) operated to make Nazi extermination technology more effective [21].

In my own work, I have shown how this preference to equate ethics to formal conformance to set procedures rather than to more complex personal ethical responsibility per se operated in the investigations of the Challenger disaster [22]. The Presidential commission recommended the institution of more procedures in response to the disaster. This recommendation is paradoxical in the face of the commission's own explicit recognition that the procedures already in place before the disaster were quite sufficient to have prevented the disaster. This paradoxical recommendation reflects, I suggest, the commission's unwillingness to avoid assigning personal responsibility.[8]

We should, therefore, avoid simplistic thinking about ethics. Winsor, for instance, cautions us against a false line of reasoning in which one can easily become entrapped [24]. It runs: 1. A disaster occurred; 2. Disasters are bad and undesirable; 3. People should behave so as to pursue the good and desirable; 4. herefore someone erred or behaved unethically. Winsor cautions, however, that we cannot assume that because a disaster occurred, someone behaved unethically.[9]

## LITERATURE REVIEW

There is a great deal of indefiniteness, ambiguity, and difference of opinon about ethics, even about such basic questions as whether ethics can be systemized at all. Though there is no natural order among these opinions, I nonetheless, have group these sources under three rough headings: systematic, professional, and academic (following G. Clark's system of professional and academic groupings, discussed below). The systematizers try to organize the varied discussions about

---

[8] D. E. Sanger discusses the commission's avoidance of assigning responsibility for the disaster [23]. He explains that one of the motivations for the out of court settlement with surviving families of the shuttle crew on condition of dropping all further claims was precisely to avert a showdown on the issue of responsibility. The commission, he says, deliberately avoided blaming individuals. To be sure, the Justice Department asserted that this settlement should not be construed as an admission of liability or negligence, (but neither, of course, should it be construed as an absolution). This avoidance by the commission of assigning personal responsibility explains, I believe, both the paradoxical conclusions of the commission and their emphasis on impersonal procedures and large-scale causes.

[9] I go one step further, however, in suggesting that, in addition to being alert against the fallacy Winsor identifies, we also should be alert against the fallacy that *that* line of reasoning is always at work in the *post facto* examination of disasters [25]. That is, though we should not always assume that someone was unethical, we also should not always assume that no one was unethical.

ethics without taking sides (though the act of systematizing itself seems to fit the academic grouping). The professional group considers ethics pragmatically, grappling with what a professional should do under given circumstances. This group is very practical, realistic, and relativistic in not taking sides on a given case and in not assuming the primacy of any particular values. The academic group considers ethics philosophically, theoretically. This group, compared to the professional, views ethics abstractly and rather absolutely, shunning relativism and arguing instead for a particular ordering of values.

## Systematic

G. Clark's *Ethics in Technical Communication: A Rhetorical Perspective* is a holistic, integrative treatment that avoids reductiveness and dichotomization [26]. Clark presents a two-fold taxonomy of perspectives of ethics, each with its own strengths and weaknesses: the professional perspective, represented by codes of ethics such as that of the Society for Technical Communication (see Brockmann and Rook [35]), and the academic perspective. Clark, however, recommends a synthesis based on classical rhetoric which avoids dichotomizing and preserves the merits of both perspectives. It is collaborative, interactive, and social constructionist. It is also highly particularized to the exigencies unique to each situation, avoiding both vague generalities and rigid specificity.

I have adopted Clark's system for this discussion for its simplicity and generality.

S. Doheny-Farina's *Ethics in Technical Communication*, Doheny-Farina reviews publications about ethics from about 1964 to 1989, divided into ethics in practice and ethics in theory [27]. The section on practice includes the professional, moral, and legal perspectives; the section on theory includes rhetorical and social constructionist views such as the civic, personalistic, and monist views of Johnson and Halloran. Doheny-Farina, at least in this essay, is not sanguine regarding the molding of an ethical communicator, taking the position that in practice all we can do is explain how ethics has been understood and practiced by others, and hope for similar conscientiousness in our students.

R. L. Johannesen's *Ethics in Human Communication* (3rd Edition) is the most comprehensive treatment of ethics regarding communication in general [28]. The ethical systems reviewed are the political; the human nature; the dialogical; the situational; and the religious, utilitarian, and legal. This work is particularly interesting for its methodology as an example of a certain frame of mind and ethic. Johannesen's method is to summarize all the different systems of ethics in a carefully even-handed, neutral, disinterested way. This method can be characterized as scientific and technical because it objectively describes an entity without interjecting preference or personal valuation. Yet the reader should recognize that this neutral, detached, relativist frame of mind has been critiqued by many critics of technology and science.

Also in Johannesen is *Social Responsibility: Ethics and New Technologies* by C. Christians, who critiques the views of Jacques Ellul on technology and technologism [29]. Also, J. V. Jensen's *Ethical Tension Points in Whistleblowing* describes the many issues involved in whistleblowing, showing that it is not the simple, straightforward matter some people take it to be [30].

## Professional

C. M. Ornatowski in *Between Efficiency and Politics: Rhetoric and Ethics in Technical Writing* explains that technical communication is not as cut-and dried as some wish it to be [31]. Ethical considerations include, for example, both protecting the environment *and* the need to maintain the viability of the organization in a competitive corporate context. Good technical communicators must address the values of the audience by giving them a communication they can accept, respect, and act upon. Similarly, good communicators realize that rhetoric is more than stylistic choices and includes the propogation and affirmation of important organization values such as technical precision and rationality.

P. Moore's *When Politeness is Fatal: Technical Communication and the Challenger Accident* illustrates how deference, politeness, and audience accommodation can complicate communication and even be unethical [32]. Moore explains that, for a great many complex social, political, and economic reasons, communication about the ice hazard at the time of the launch of Challenger was couched in very conciliatory, inoffensive, ambiguous language. The communicators' scrupulous accommodation to the values and interests of the managerial audience thus worked to the disastrous detriment of the non-managerial audience, specifically the Challenger crew. Moore also notes that even the supposedly "hard," objective data of a technical report are socially constructed even before the reception of the communication, profoundly implicating the audience in the ethics of communications.

D. L. Sturges's *Overcoming the Ethical Dilemma: Communication Decisions in the Ethic Ecosystem* takes an algorithmic approach to ethics using a decision-tree scheme [33]. This tree-like decision process is based on simple Yes-No answers to specific questions. This approach has a distinct impersonal, proceduralist flavor in making the achievement of goals paramount and in controlling the bases for judging actions. Sturges does not address the question of whether ethical decisions can satisfactorily be reduced to this simple algorithm or procedure.

T. M. Sawyer's *The Argument About Ethics, Fairness, or Right and Wrong* contends that the decision-making processes of the appellate court system can serve as a useful model of ethical decision making [34]. This approach begins with an exposition of the facts, then reasons from categorical syllogisms qualified by exceptions to a majority opinion (which itself can be qualified by a dissenting opinion.) Sawyer has found that his students usually find this systematic approach to be helpful in shaping effective, persuasive arguments on ethical issues.

*Technical Communication and Ethics*, an anthology published by the Society for Technical Communication and edited by Brockman and Rook, can be considered the official statement on ethics for technical communicators, if any work has the right to be so called [35]. Its reprints cover a range of topics including the basis for ethics; the influence of disciplinary context; the connections between ethics and rhetoric; the teaching of ethics; and a brief history of ethics in technical communication. It takes a stance toward the topic: appreciation of the ethical complexity of actual situations, careful neutrality, and practicability (with some exceptions such as Rubens's "Reinventing the Wheel?" [36]). An indication of this neutrality is that it takes no position itself, instead offering an objective presentation and allowing readers to decide for themselves. We should remember, however, that for critics such as Bizzell and Katz this neutral stance is already itself an ethical position. Such critics emphasize that neither the communicator nor the "observer" of communicators can absolutely elevate himself or herself above his or her own ethical responsibilities and engagement. It should also be noted that the more abstract, theoretical and judgment-oriented academic perspective is not present in this work.

W. J. Buchholz in *Deciphering Professional Codes of Ethics* explores the difficulties with many professional codes of ethics [20]. First, they have problems with generalizing; many real events are so specific to their unique situations that they cannot be generalized to other events. Second, many codes are "encoded," couched in such arcane, specialized terminology that they are only tenuously connected to any commonsense notion of ethics. Their, attitudes toward either the public or truth can be unrealistic, unwarranted, or difficult to operationalize.

Buchholz cites the STC Code for Communicators as an example of problematic assumptions about knowledge and truth. These are precisely the assumptions I discussed early regarding positivism: that truth is entirely "objective," "separable from the individual," "verifiable," "discrete," and "absolute" [20, p. 68]. Instead of this "absolutist" stance toward truth, Buchholz argues for a more conditional, relativistic, and contingent stance he calls "humanistic" [20, p. 68]. This humanistic stance, rather than focusing on the Truth per se, focuses instead on the truthfulness of the person, a subtle but very significant distinction (see my similar distinction between impersonal procedures and personal judgment regarding Challenger [37].)

Buchholz offers six principles for shaping codes of ethics into systems for realistic guidance and usefulness. Among these are classifying troublesome practices or principles; indicating the standing of the law on an issue; evaluating perennial knotty problems; and offering explicit interpretations with historical and philosophical perspectives.

H. Sachs's *Ethics and the Technical Communicator* views communication as best which is the impersonal, non-subjectivity relaying of information [38]. This position, it is important to note, differs from the positions of many of the critics cited here in asserting the ethical responsibility of the technical communicator *not* to shape the communication against the organization's intentions. Managers and decision-makers can rightfully expect their own authority to be recognized, the good exercise of which requires accurate, clear information. Any alteration would undermine their rightful authority and in effect prevent proper exercise of their own responsibilities. This position relieves the technical communicator of the burden of very broad ethical responsibility and of taking on the responsibilities properly attached to others, the sender and the receiver of the communication.

## Academic

P. T. Durbin's *Technology and Responsibility* considers the philosophical connections between technology and responsibility [39]. I particularly recommend in it W. H. *Vanderburg's Technique and Responsibility: Think Globally, Act Locally, According to Jacques Ellul* for its careful summarization of Ellul's views on technology and ethics [40]. Vanderburg urges against an instance-by-instance critique of technology by various ethical systems, offering instead an ethical criticism of the whole system of technologies and techniques that constitutes our interconnected social milieu.

J. C. Pitt's *The Autonomy of Technology*, also in Durbin, expresses similar reservations about the basic idea of autonomous technology [41]. To my mind, though, the question is less whether technology or technologism *is* autonomous in an absolute, essentialist sense but more whether we choose to take it as so.

J. P. Zappen's *The Discourse Community in Scientific and Technical Communication: Institutional and Social Views* is relevant to ethics indirectly [42]. Zappen argues that viewing technology and science within the frame of reference of particular disciplines is neither entirely adequate nor warranted because the "discourse community" in which technical or scientific writing occurs is not limited to these "institutions." Rather communication also occurs within a larger, more general "social" community that embraces both other disciplines and the general public. The technical communicator therefore is never disengaged from his or her participation in and responsibilities to both the social and institutional communities. Zappen thus supports the view that confining considerations of ethics solely to particular technical disciplines is neither warranted nor fully adequate.

M. Kremers's *In Teaching Ethical Thinking in a Technical Writing Course* draws the important distinction that controversy and disagreement are good in themselves as indications that decisions are being examined from a variety of perspectives and values—it is all too easy to think simplistically that disagreement is disagreeable and should be avoided [43]. Instead, Kremers opposes having students take a stance of careful neutrality and relativism, for to do so fosters the avoidance of actually making ethical judgments and decisions.

S. Katz's *The Ethic of Expediency: Classical Rhetoric, Technology, and the Holocaust* uses the example of a Nazi technical document urging improvements in vans used to kill Jews and other undesirables to show that technical perfection, whether of a technology and of a technical communication, embodies an ethic of expediency that can be dangerous because it distances humans from other humans [21]. This potential danger is inherent in technology itself, Katz contends.[10] This principle of expediency is dangerous when it is the exclusive ethical ground of a communication, because it fosters such horrifying objectivizations as the Nazi memo in which people are referred to as "the load" and "pieces."

K. Possin takes issue with Sawyer's approach in his *Ethical Argumentation* (see Professional group above) [44]. Sawyer's approach, Possin says, is basically sophistical, arriving at different conclusions depending on which initial premises one wishes to adopt. Possin offers, instead, argument by analogy as a more valid, reasonable, satisfying (in that its conclusions are binding), practicable, and simple approach to ethics based on generalizing from clear prior cases. Assuming that clear prior cases can be found, that the determination of the rightness or wrongness in that case is uniformly agreed on, and that the present case is truly analogous to the prior case, then the conclusion is absolutely clear and binding for to conclude otherwise would show the arguer to be inconsistent or irrational.

N. Johnson in *Ethos and the Aims of Rhetoric* (like C. R. Miller in the chapter on social constructionism) challenges the assumed moral neutrality of technology and science, and, like Bizzell, challenges the pedagogical stance of neutral, detachment, and radical relativism [45]. Johnson argues from classical and traditional rhetoric that communication is inescapably, intrinsically ethical in embodying values and interests. She also argues for the necessity of relying on the wisdom and judgment on the communicator as a person.[11] Ethical responsibility lies fundamentally in the person and is effected socially through individual persons persuading others.

---

[10]Ellul says the same, though using his term variously translated as technicism or technologism. He defines technologism [his *technique*] as "the totality of methods, rationally arrived at and having as its goal absolute *efficiency* [emphasis mine] (for a given state of development) in every field of human activity" [3, p. xxv].

[11]As I explain in *The Lessons of the Challenger Investigations*, one of the findings of the Presidential commission investigating the Challenger disaster was that impersonal systems of procedures can never substitute for personal judgment, and should never be expected to do so [22].

B. Barton and M. Barton argue along lines similar to Johnson in their *Ethos, Persona, and Role Confusion in Engineering: Toward a Pedagogy for Technical Discourse* [46]. They argue that a paradigm shift has occurred in communication away from the formalist approach emphasizing the precision of the text and the passivity of the transmitter and receiver of the classic communication model, toward a social constructionist approach recognizing the constructed, contingent nature of knowledge and the creative power of the communicator as person. They also criticize the impersonalness often recommended for technical communication as not the absence of personal persona but only itself an odd persona, one which futilely tries to deny itself.

E. Levinas, though not a technical communicator, seems nonetheless relevant to our discussion in addressing many of the concerns critics cited here have raised [47]. This post-modernist continental theorist has received a good deal of attention from J. Derrida and others since the late 1970s. His approach to ethics rejects objectifying, emphasizes particularlity and situational specificity, and is neither relativistic nor absolutist.

Levinas holds that morality begins with our awareness of and response to what he calls "the face of the other person." This "other" is a particular (rather than generic) other person, whose otherness makes him or her fundamentally different from ourselves. This uniqueness and difference is embodied in "the face," the usual locus of our distinguishing one person both from ourselves and from all others. The essential particularity of this distinctly-other person opposes any inclination we might have to generalize, to lay down and act out of general rules of conduct, because the effect on feelings, beliefs, and wishes of that particular person are inherently unknown (at least initially) and cannot be assumed or fully anticipated. In addition, the other's ability to respond, to criticize, and to disagree elementally opposes our inclination to objectivize him or her.

An example of the moral, emotional, and rhetorical force of the particularizing of ethics that Levinas advocates is the Vietnam Memorial in Washington. Instead of the generic figures seen in other monuments (such as the statues of Vietnam soldiers nearby), the Vietnam Memorial presents each individual name before the viewer, who in effect *faces* the death of each one of these individuals while seeing his or her actual human face optically reflected in the polished granite.

A realistic instance of how a Levinasian perspective could influence technical communication can be drawn from the investigations of the Challenger disaster. One of the findings of the Presidential commission and the Congressional committee was that the shuttle astronauts, those most directly affected by decisions about flightworthiness and the significance of O-ring charring, had not been included in the decision-making process. Thus the impersonal process proceeded without having to *face* directly the people affected by those decisions. Following the recommendations of the investigators, astronauts now are included in such decisions, a real face before the decision makers.

More importantly, Levinas explains that we must continually face the unremitting task of ethical examination. We might say, paraphrasing Socrates, that the ethically unexamined act is not worth doing. Indeed, one of the reasons Socrates was so chronically troublesome for Athenians was that he specifically rejected easy answers and pat judgments.

## REPRINTED ARTICLES

### M. H. Markel

In *A Basic Unit on Ethics for Technical Communicators*, Markel discusses the meaning of ethics, reviews recent scholarship, presents a brief history of ethics relating to communication, summarizes basic approaches to ethics in technical communication, and applies these approaches to a realistic, pedagogical case [19]. He recommends a case approach to teaching ethics in order to avoid excessive abstractness while demonstrating the great complexity of real-world ethical dilemmas. Only particular instances can embody the complexity and interrelatedness of the various values operative in any ethical dilemma, Markel holds. Because of this crucial uniqueness of each particular episode, he finds codes of conduct to be of little relevance or usefulness: they are simply too general and vague by their very nature. "Because codes have to be flexible enough to cover a wide variety of situations, they tend to be so vague that it is virtually impossible to say that a person in fact violated one of their principles" [19, pp. 34-35]. Markel's purpose is pedagogical. Markel has his students discuss the employer-employee relationship as a social contract entailing specific rights and obligations both for the employee and the employer such as competence, diligence, honesty, and confidentiality. The employee also has obligations to the public which Markel explains through contract theory, due-care theory, and strict-liability theory. The employees obligation to the environment is to preserve ecosystems and maintain diversity of species. All these obligations must be considered and weighed against each other in coming to a decision about how to behave ethically.

Thus Markel's approach is highly practical. It does not, for instance, pursue theory abstractly or absolutely but plays interests against each other relativistically and pragmatically. This approach then, from the scheme of Clark, represents the professional rather than the academic approach.

### D. Sullivan

Sullivan's *Political-Ethical Implications of Defining Technical Communication as a Practice* is an important statement in its scope; its sensitive discussion of classical rhetoric; and its explicit treatment of the social context of industrial, technological, and scientific activities [48]. It is of an academic sort (per Clark) in being theoretical but is nonetheless practical and non-idealistic. While some readers might initially be troubled by the "political" in the title, the earnest reader will be rewarded by a articulate discussion of technical communication of a distinctly humanistic sort.

Sullivan's principal point is to criticize the conventional understanding of technical communication instruction as the passing on of a set of impersonal skills and the mastering of conventional forms. Sullivan advocates instead an understanding of technical communication as an intensely social practice involving judgment, empowerment, politics, and ethics. His main objection to technical communication as only skills is that it grants an unwarranted, almost autocratic authority to technology. It also perpetuates the disempowerment of the communicator as a worker who slavishly acquiesces to the forces of production and technology.

Sullivan draws from the history of rhetoric itself (citing Johnson's Three Nineteenth-Century Rhetoricians: The Humanist Alternative to Rhetoric as Skills Management [49]) in looking to early American rhetoricians Theremin, Day, and Hope as well as to Aristotle for alternative views of rhetoric emphasizing *praxis* and social action to the mechanistic, skills view.

Sullivan presents an alternative, humanistic approach to technical communication that encourages the critical examination of forms, genres, and lines of argument while it contextualizes both technology and technical communication within the larger society. Responsible social action requires the exercise of "practical wisdom or prudence (*phronesis*)," and such prudential judgment should be cultivated in our classrooms. The effects of adopting this alternative view would be to "bring rhetoric out of the amoral realm of technique into the world of ethics and politics" [politics in the historical sense of civics] and to act "in the citizen's role rather than in the worker's role" [48, p. 378].

A basic assumption underlies Sullivan, however, with which some readers might not readily agree and which I myself have questioned (see Chapter One, Humanism and Technical Communication). He assumes an agonistic relationship, a relationship of radical opposition, between technology and humanism. This agonism is reflected naturally in Sullivan's distinctly suspicious, adversarial view of technology. And just as naturally it perpetuates a polarizing duality which makes it doubly difficult to define a middle ground which neither categorically rejects nor categorically embraces notions embodied in the opposing polar extremes. Sullivan, to be sure, himself begins to articulate such a middle ground, at least in his recognition of the need to accommodate the competing demands of skills competency *and* social action as well as in his example from his own teaching in which mastery of technical skills comes before the critical and selective implementation of those skills.

## REFERENCES

1. A. Zucker, D. Borchert, D. Stewart (eds.), *Medical Ethics: A Reader*, Prentice Hall, Englewood Cliffs, New Jersey, 1992.
2. P. Singer (ed.), *A Companion to Ethics*, Basil Blackwell, Oxford, United Kingdom, 1991.
3. J. Ellul, *The Technological Society*, Knopf, New York, 1964.
4. R. M. Weaver, *Language is Sermonic*, R. L. Johannesen, R. Strickland, and R. T. Eubanks (eds.), Louisiana State University Press, Baton Rouge, Louisiana, 1970.
5. S. V. Monsma, *Responsible Technology: A Christian Perspective*, For Calvin Center for Christian Scholarship, Eerdmans, Grand Rapids, Michigan, 1986.
6. A. G. Gross, Discourse on Method: The Rhetorical Analysis of Scientific Texts, *Pre/Text, 9*:3-4, pp. 169-185, 1988.
7. D. N. Dobrin, Is Technical Writing Particularly Objective? *College English, 47*:3, pp. 237-251, 1985.
8. C. Gilligan, *In A Different Voice*, Harvard University Press, Cambridge, Massachusetts, 1982.
9. N. Noddings, *Caring: A Feminine Approach to Ethics & Moral Education*, University of California, Berkeley, California, 1984.
10. M. J. Larrabee, *An Ethic of Care*, Routledge, New York, 1993.
11. K. Burke, *Language as Symbolic Action*, University of California Press, Berkeley, California, 1966.
12. S. M. Halloran, Eloquence in a Technological Society, *Central States Speech Journal, 29*, pp. 221-227, 1978.
13. P. Bizzell, Cognition, Convention, and Certainty: What We Need to Know About Writing, *Pre/Text, 3*:3, pp. 213-243, 1982.
14. P. Bizzell and Bruce Herzberg, *The Rhetorical Tradition*, Bedford Books of St. Martin's Press, Boston, Massachusetts, 1990.
15. C. R. Miller, A Humanistic Rationale for Technical Writing, *College English, 40*:6, pp. 610-617, 1979.
16. T. Trainer, *The Nature of Morality*, Avebury Press, Aldershot, England, 1991.
17. F. L. Wright, *An Autobiography*, Horizon Press, New York, 1932.
18. United States, Presidential Commission on the Space Shuttle Challenger Accident, *Report to the President by the Presidential Commission on the Space Shuttle Challenger Accident*, 86-16083, GPO, Washington, D. C., 1986.
19. M. Markel, A Basic Unit on Ethics for Technical Communicators, *Journal of Technical Writing and Communication, 21*:4, pp. 327-350, 1991.
20. W. J. Buchholz, Deciphering Professional Codes of Ethics, *IEEE Transactions on Professional Communication, 32*:2, pp. 62-68, 1989.
21. S. B. Katz, The Ethic of Expediency: Classical Rhetoric, Technology, and the Holocaust, *College English, 54*:3, pp. 255-275, 1992.
22. P. M. Dombrowski, The Lessons of the Challenger Investigations, *IEEE Transactions on Professional Communication, 34*:4, pp. 211-216, 1991.
23. D. E. Sanger, Shuttle Accords Avert a Showdown, *The New York Times*, sect. D, p. 3, col. 1, January 2, 1987.

24. D. A. Winsor, The Construction of Knowledge in Organizations: Asking the Right Questions about the Challenger, *Journal of Business and Technical Communication,* 4, pp. 7-20, 1990.

25. P. M. Dombrowski, A Comment on "The Construction of Knowledge in Organizations: Asking the Right Questions about the Challenger," *Journal of Business and Technical Communication, 6*:1, pp. 123-127, 1992.

26. G. Clark, Ethics in Technical Communication: A Rhetorical Perspective, *IEEE Transactions on Professional Communication, 30*:3, pp. 190-195, 1987.

27. S. Doheny-Farina, Ethics and Technical Communication, in *Technical and Business Communication: Bibliographic Essays for Teachers and Corporate Trainers*, C. H. Sides (ed.), Co-published by National Council of Teachers of English and Society for Technical Communication, Urbana, Illinois, 1989.

28. R. L. Johannesen (ed.), *Ethics in Human Communication,* (3rd Edition) Waveland Press, Prospect Heights, Illinois, 1990.

29. C. Christians, Social Responsibility: Ethics and New Technologies, in *Ethics in Human Communication,* (3rd Edition) R. L. Johannesen (ed.), Waveland Press, Prospect Heights, Illinois, 1990.

30. J. V. Jensen, Ethical Tension Points in Whistleblowing, in *Ethics in Human Communication,* (3rd Edition) R. L. Johannesen (ed.), Waveland Press, Prospect Heights, Illinois, 1990.

31. C. M. Ornatowski, Between Efficiency and Politics: Rhetoric and Ethics in Technical Writing, *Technical Communication Quarterly, 1*:1, pp. 91-103, 1992.

32. P. Moore, When Politeness is Fatal: Technical Communication and the Challenger Accident, *Journal of Business and Technical Communication, 6*:3, pp. 262-292, 1992.

33. D. L. Sturges, Overcoming the Ethical Dilemma: Communication Decisions in the Ethic Ecosystem, *IEEE Transactions on Professional Communication, 35*:1, pp. 44-50, 1992.

34. T. M. Sawyer, The Argument About Ethics, Fairness, or Right and Wrong, *Journal of Technical Writing and Communication, 18*:4, pp. 367-375, 1988.

35. R. J. Brockmann and Fern Rook (eds.), *Technical Communication and Ethics*, Society for Technical Communication, Arlington, Virginia, 1989.

36. P. M. Rubens, Re-inventing the Wheel: Ethics for Technical Communicators, *Journal of Technical Writing and Communication, 11*, pp. 329-339, 1981.

37. P. M. Dombrowski, *People or Procedures? The Ethical Lessons of the Challenger Disaster*, 1989 Conference of The Society for the Social Studies of Science, University of California, Irvine, 1989.

38. H. Sachs, Ethics and the Technical Communicator, *Technical Communication, 27*, pp. 7-10, 1980.

39. P. T. Durbin (ed.), *Technology and Responsibility*, D. Reidel, Dordrecht, The Netherlands, 1987.

40. W. H. Vanderburg, Technique and Responsibility: Think Globally, Act Locally, According to Jacques Ellul, in *Technology and Responsibility,* P. T. Durbin (ed.), D. Reidel, Dordrecht, The Netherlands, 1987.

41. J. C. Pitt, The Autonomy of Technology, in *Technology and Responsibility,* P. T. Durbin (ed.), D. Reidel, Dordrecht, The Netherlands, 1987.

42. J. P. Zappen, The Discourse Community in Scientific and Technical Communication: Institutional and Social Views, *Journal of Technical Writing and Communication, 19*:1, pp. 1-11, 1989.

43. M. Kremers, Teaching Ethical Thinking in a Technical Writing Course, *IEEE Transactions on Professional Communication, 32*:2, pp. 58-61, 1989.

44. K. Possin, Ethical Argumentation, *Journal of Technical Writing and Communication, 21*:1, pp. 65- 72, 1991.

45. N. Johnson, Ethos and the Aims of Rhetoric, in *Essays on Classical Rhetoric and Modern Discourse*, R. J. Connors, L. S. Ede, and A. Lunsford (eds.), Southern Illinois University Press, Carbondale, Illinois, 1984.

46. B. Barton and Marthalee Barton, Ethos, Persona, and Role Confusion in Engineering: Toward a Pedagogy for Technical Discourse, *Proceedings of the Conference on College Composition and Communication*, ERIC Document 229 782, 1981.

47. E. Levinas, *Collected Philosophical Papers*, A. Lingis (trans.), Martinus Nijhoff, Dordrecht, The Netherlands, 1987.

48. D. L. Sullivan, Political-Ethical Implications of Defining Technical Communication as a Practice, *Journal of Advanced Composition, 10*, pp. 375-386, 1990.

49. N. Johnson, Three Nineteenth-Century Rhetoricians: The Humanist Alternative to Rhetoric as Skills Management, in *The Rhetorical Tradition and Modern Writing*, J. J. Murphy (ed.), Modern Language Association, New York, 1982.

# A BASIC UNIT ON ETHICS FOR
# TECHNICAL COMMUNICATORS

## *Mike Markel*

The technical communication class began routinely enough, with the organizational patterns of comparison and contrast discussions: whole-by-whole and part-by-part. The students were already at work on their report assignment which for many of them was a feasibility study involving comparison and contrast of options.

The fun started when a student offered a comment. She said that at her last co-op assignment, she had worked for a consulting company that recommended computer hardware and software to businesses, particularly tax-accounting firms. The recommendations took the form of feasibility studies in which the author compared and contrasted the different available hardware and software packages. Her supervisor instructed the employees to structure the reports to emphasize how hard the consulting company had worked in analyzing the client's situation: the recommended option was always discussed last in a whole-by-whole pattern.

"There's nothing wrong with that," I responded. Her company has every right to get fair compensation for its service. After an uncomfortably long pause she responded that the reports made it seem as if she had worked a lot harder than she actually had. Most of the projects the company took on were quite similar; after a few months on the job, she had developed effective boilerplate for the routine projects. She said she didn't think it was right to charge a lot of money for what was in fact a simple assignment.

"But if her recommendations are good, what is the harm?" another student asked. "The client is getting what it wants, and the consulting company is making a good profit. Everyone wins." "Yet what if the client finds out that it paid thousands of dollars for a report that took a co-op student an hour and a half at the word processor? The client would never do business with the consulting company again, and it would do everything possible to spread the word about its bad experience."

"So the consulting company's task is to make sure the client never finds out how quickly one of the feasibility reports can be turned out.

It was becoming painfully clear to me that in the last twenty minutes we had entered deep into territory for which the textbook provided no reliable map. (By the way, I wrote the textbook.) Before the hour was out, we had gone much further.

Isn't overcharging for the report a form of stealing from the client and indirectly a cause of inflation for everyone? Yes, but all consulting companies make a huge profit on the simple cases to finance the complicated ones. Therefore, doesn't this company have to do it to stay in business? Isn't it their right—no, their obligation—to provide employment for their workers? Okay, but if you do good work at a fair price, eventually you will prosper and achieve the same ends ethically. But can't a company set its own fees? Of course, but it's a question of ethical representation to the client: if you charge a competitive hourly fee and inflate the hours, that's unethical; if you are honest about the number of hours the job will take, you're giving the client the right to make up his or her own mind.

The discussion ended (it didn't conclude) at the hour. In thinking about what had occurred in the classroom, I realized that we had raised some of the major issues pertaining to customer-client relationships. But there were many more questions relating to other aspects of technical communication, such as the employee's rights and obligations to his or her employer and to the general public, that needed to be addressed.

This essay describes a basic unit for technical communicators—both professional communicators and professionals who communicate—and offers some suggestions on how to go about teaching the unit.

## RECENT SCHOLARSHIP ON ETHICS

In recent years, the amount of scholarship relating ethics to applied communication has increased dramatically. The September 1990 issue of the *Bulletin of the Association for Business Communication* is devoted to the relationship between ethics and the teaching and practice of on-the-job writing [1]. A 1985 issue of the *Journal of Business Communication* contains several provocative essays on ethics [2]; the STC anthology *Technology Communication and Ethics* contains some sixteen essays [3]. A special issue of *IEEE Transactions on Professional Communications is* devoted to legal and ethical considerations in technical communications [4]. And the STC/NCTE joint publication, *Technical and Business Communication: Bibliographic Essays for Teachers and Corporate Trainers* contains a comprehensive essay on ethics [5].

Many teachers of technical communication have offered suggestions about ways to teach ethics in their courses. For example, Rentz and Debs argue that the values and assumptions that we communicate through our diction are the most appropriate focus for instruction in ethics [6]. Krohn recommends a general semantics approach, focusing on such questions as metaphors, level of abstractions, and loaded words [7]. Kremers teaches ethical thinking by having students research and write about the Strategic Defense Initiative, an ethically sensitive project [8]. Sachs presents nine case problems to be used in a technical communication course to stimulate thinking about ethics [10].

Although each of these approaches is useful, my classroom experience and my discussions with other teachers have led me to conclude that many students lack a basic understanding of how to think through ethical problems. Without an awareness of the fundamental principles used to navigate tricky ethical waters, they tend to rely on the quick and often superficial ethical thinking represented in my re-creation of the day's discussion in class.

Of course, numerous scholars have sketched in the outlines of different ethical theories. For example, Johannesen provides a set of perspectives, such as legal, religious, political, and psychological, used to interest students in the ethical issues inherent in all discourse [11]. Wicclair and Farkas briefly describe three types of ethical principles—goal-based, duty-based, and rights-based—that can guide students through several case studies [12]. Golen et al., in a report from the Association for Business Communication's Teaching Methodology and Concepts Committee subcommittee on the teaching of ethics, present three basic principles: social utility, general law (Kant's categorical imperative), and long-range utility [13]. In addition, they list eight "common sense" principles of ethics (including candor, social harmony, fidelity, consistency of word and act, and maintaining confidence) originally articulated by John C. Condon [14].

Yet these discussions are not fully adequate. Some are too narrow in that they restrict themselves to the ethical questions raised at the point of the communication itself: misrepresentation of information, plagiarism, and the like. Although these ethical questions are critical and must be addressed, our students also will be confronted with a wider range of ethical issues that lie outside of and are antecedent to the issues involved in the process of communication. While it could be argued that teaching these broader ethical issues should not be the sole or even the primary responsibility of the writing teacher, clearly they are not being addressed systematically in the technical disciplines themselves.

If some of the existing discussions of ethical principles related to technical communication are too narrow, others are too broad; they outline Kant's categorical imperative or the general-utility principle, but they don't connect the theory to the kinds of practical problems the students will confront when they begin their careers. These studies do not go into sufficient detail to enable students to understand how the different theories conflict with each other on the job. Although virtually every such discussion includes a caveat that it will not enable the reader to reach a simple solution to every ethical problem, what is lacking is a discussion that will show in some detail how different goals come into conflict in the presence of a particularly complex ethical dilemma. Yet it is the complex ethical dilemma that students need to be able to confront successfully.

In *Reinventing the Wheel?: Ethics for Technical Communicators*, Philip Rubens urges us to survey the existing work on ethics in fields related to or similar to technical communications as a first step in devising our own guidelines [15]. Rubens reviews the literature in journalistic ethics. Certainly, this is sensible, because technical communication, especially in its rhetorical theory, is very close to journalism. In this essay I would like to focus on a different literature: business and professional ethics. This approach examines the role of the communicator within a corporate or institutional context. Most technical communicators and other professionals who communicate technical information work in government, business, and industry. They therefore inhabit the world of capitalistic motivations, with its emphases on reducing R&D and production costs by using the least expensive materials and processes, on getting products into the marketplace quickly, and on communicating positive information to the public. These emphases contribute to many of the intense pressures to act unethically. Studies in a number of different industries suggest that some 60 to 70 percent of managers feel under some pressure to compromise their own code of ethics on the job [16, p. 7]. Neither technical communicators nor other professionals are exempt from this pressure.

This article shows that the literature of business and professional ethics can form a bridge between the broad discussions of general ethical principles and the narrow discussions of plagiarism and the like. This ethics scholarship addresses the needs of technical communicators and other professionals while retaining a sufficiently comprehensive focus on them as workers within an organizational context.

## ETHICS IN CLASSICAL RHETORICAL LITERATURE

Any discussion of ethics and writing should begin with an introduction to the subject as it was explored in the classical (especially the Greek) rhetorical literature. This subject has, of course, been treated in numerous book length studies (see, especially, Kennedy [17, 18] and Vickers [19]). In addition, the journals *Rhetoric Review, Quarterly Journal of Speech,* and *Philosophy and Rhetoric* explore the subject regularly. In the interest of brevity, I will merely sketch in the outlines of the subject.

### Philosophy and Rhetoric

The relationship between philosophy and rhetoric, to use the terms from the classical literature, is extremely complex and ultimately unresolved. Plato set the terms of the debate in the *Gorgias* and the *Phaedrus* with his well known attack on rhetoric, which he claims is not an art but merely a form of verbal manipulation, a kind of flattery. Rhetoric is a technique, not a subject. It has no substance and is therefore unrelated to truth, which can be discovered only through philosophy's dialectic method. Plato's definition of rhetoric is thus the prototype and archetype of the modern pejorative use of the term. Opposing Plato were the Sophists, who believed that rhetoric was an indispensable tool in making practical political decisions. Forerunners of modern western ideas of jurisprudence, the Sophists taught that rhetoric enabled both sides in a dispute to present their cases persuasively and therefore enabled the jury to reach a reasonable verdict. Sophists were less concerned than Plato with the notion of truth. For example, Protagoras, a leading Sophist, believed that truth was unknowable and perhaps nonexistent. As Kennedy points out, Plato and the Sophists represented not merely two different views of the nature and aims of rhetoric, but two different views of reality: Plato believed that truth exists and is knowable only through philosophy, whereas the Sophists believed that, deprived of a clear knowledge of truth, people must use rhetoric skillfully to achieve wise and just ends [18]. What defined wisdom and justice, to the Sophists, were the prevailing cultures and institutions. As Vickers demonstrates, Plato's attacks on rhetoric are themselves masterful examples of rhetoric whose goal is to lead the reader to truth [19]. Therefore, if they persuade the reader, they negate their own thesis.

The greatest rhetorician, of course, was Aristotle, who refined and extended the principles of rhetoric. Aristotle taught that rhetoric is the counterpart of dialectic Both are arts used by common people, not only by specialists. Like the Sophists, Aristotle argued that rhetoric is useful in the pursuit of truth, because it enables the jurors or listeners to understand the issues being debated. (However, Yoos argues persuasively that even Aristotle's conception of *ethos* is too far removed from ethics, focusing on the listener's emotional response to the message rather than on the ethical appeal of the message itself [20].) In response to Plato's argument that rhetoric can be misused by unscrupulous people, Aristotle responded that the same charge can be leveled against everything except virtue itself.

The debate about philosophy and rhetoric is a fascinating intellectual puzzle that is clearly at the heart of the questions my students were grappling with. Is rhetoric merely a tool for victimizing the unsophisticated, or is it a means of realizing justice and truth? The question is as current as any political campaign. Yet despite the centrality of the debate, it is not sufficient for a clear understanding of the relationship between ethics and modern technical communication; the classical rhetoricians did not explicitly treat many of the specific dilemmas that occupy the modern technical communicator.

## ETHICS IN TECHNICAL COMMUNICATION

Modern technical communicators—a term that encompasses not only people called technical writers and editors but also all people who communicate technical information as part of their jobs—will probably be working in business, industry, or government, and will therefore have to make personnel decisions involving hiring, salaries, promotions, and dismissals within the organization. They will have to make decisions about how to treat clients, customers, and other organizations. They will have to make decisions about how their organizations deal with government regulatory bodies. And they will have to make decisions about how the actions of their organizations affect the environment.

To help the technical-communication instructor present a basic unit on ethics and technical communication, I have divided this essay into five sections:

1. A brief definition of ethics and an explanation of some of the standards commonly used to reach decisions on ethical questions.
2. An explanation of the employee's three basic obligations: to the employer, the public, and the environment.
3. A discussion of some ways to analyze the common dilemmas that a technical communicator faces routinely, concerning such topics as plagiarism, trade secrets, the fair use of language and visuals in product information and advertising, and whistleblowing.
4. A brief discussion of the role of the code of conduct: its strengths and limitations as a mechanism for defining and enforcing ethical behavior in the workplace.
5. A case study showing the dilemma faced by a technical writer, followed by an explanation of how to use the case in the classroom.

## A BRIEF INTRODUCTION TO ETHICAL THINKING

The list of ethical issues that have been the focus of critical scrutiny in recent years is lengthy. The mere mention of names is enough to suggest the range of persons whose business and personal affairs have been the subject of ethical investigation: televangelists Jim and Tammy Bakker and Jimmy Swaggart, hotel tycoon Leona Helmsley, Secretary Samuel Pierce of the Department of Housing and Urban Development, Congressmen Jim Wright and Barney Frank, Canadian sprinter Ben Johnson, financiers Ivan Boesky and Michael Milken.

Other familiar phrases recall ethical controversies located within a more complex corporate or organizational context: Iran-contra and the question of lying to Congress; Exxon's response to the Alaskan oil spill; sanctions against organizations operating in South Africa; Union Carbide's handling of the tragedy in Bhopal, India; the Challenger explosion and the defective O-rings. The rights of the unborn versus the rights of the pregnant woman. The rights of the minority applicant versus the rights of the non-minority applicant. The list goes on and on.

### But What Exactly Is Ethics?

Naturally, there are a number of definitions. Some people equate ethical conduct and legal conduct; if an act is legal, it is ethical. Most people, however, believe that ethical standards are more demanding than legal standards. It is perfectly legal, for example, to try to sell an expensive life insurance policy to an impoverished elderly person who has no dependents and therefore no need for such a policy, yet many people would see such an act as unethical or bordering on the unethical. Some see ethical behavior as merely telling the truth or showing respect for others.

When businesspersons were asked to give their definition of ethics, half of them answered "What my feelings tell me is right." One quarter of the respondents answered, "What is in accord with my religious beliefs." And most of the rest of the businesspersons said, "What conforms to the golden rule" [21]. Rather than reducing ethics to a formula, most philosophers prefer to define it as a field of enquiry; for them, ethics is the study of the principles of conduct that apply to an individual or a group.

## Three Standards of Ethics

What are the basic principles used to think through an ethical problem? Ethicist Manuel G. Velasquez outlines three kinds of moral standards useful in confronting ethical dilemmas: rights, justice, and utility [22].

The standard of rights focuses on the basic needs and welfare of particular individuals. If a company has agreed to provide continuing employment to its workers, the standard of rights requires that the company either keep the plant open or provide adequate job training and placement services.

The standard of justice asks this question: how should the positive and negative effects of an action or policy be distributed among a group? For example, the standard of justice would suggest that the expense of maintaining a highway be borne primarily by persons who use that highway. However, since everyone benefits from the highway, it is just that general funds also be used for maintenance.

The standard of utility asks this question: What will be the effects, both positive and negative, of a particular action or policy on the general public? If, for example, a company is considering closing a plant, the standard of utility requires that the company's managers consider not only any savings they will reap from shutting down the plant, but also the financial hardships on the unemployed workers and the economic impact on the rest of the community.

Although it is best to think about the implications of any serious act in terms of all three standards, often there will be a conflict among them. For instance, from the point of view of utility, no-fault car insurance laws—which stipulate that people may not sue other people for damages under a certain dollar amount—are a good idea because they reduce the number of nuisance suits that clog up the court system and increase everyone's insurance costs. However, no-fault car insurance laws seem to violate the standard of justice; it is unfair that the insurance company of a driver who is not at fault have to pay for the repairs.

In the cases of conflicts among the three standards, the standard of rights is generally considered the most important and the standard of justice the second most important. The complicating factor, however, is that the three standards cannot simply be ranked in terms of importance as a means of solving all ethical problems. If an action or policy were to have a great effect according to the standard of utility, the least important standard, this factor might outweigh any effects according to the standard of rights. For instance, if the power company has to cross your property to repair a transformer on a utility pole, the standard of utility (the need to get power to all the people affected by the problem) takes precedence over the standard of rights (your right to private property). Of course, the power company is obligated to respect your rights, insofar as possible, by explaining what it wants to do, trying to accommodate your schedule, and repairing any damage it might do to your property. Ethical problems are difficult to resolve precisely because there are no rules to determine when one standard outweighs another. In the example of the transformer, how many customers have to be deprived of their power before the power company is morally entitled to violate the property owner's right of privacy? Is it 1,000? 500? Ten? Only one?

Most people will not debate the conflict among rights, justice, and utility when they are confronted by a serious ethical dilemma; instead, they will do what they think is right. Perhaps this is good news. However, the quality of ethical thinking varies dramatically from one person to another, and the consequences of superficial ethical thinking can be profound. For these reasons, ethicists have described a general set of principles that can help a person organize his or her thinking about the role of ethics within an organizational context. This set of principles is a web of rights and obligations that connect an employee, an organization, and the world in which it is situated.

## THE EMPLOYMENT CONTRACT

In exchange for his or her labor, the employee, for example, enjoys three basic rights—generally acknowledged in the modern era to be fair wages, safe and healthy working conditions, and due process in the handling of such matters as promotions, salary increases, and firing. Although there is still serious debate about employee rights, such as the freedom from surreptitious surveillance and unreasonable search in the case of drug investigations, the question almost always concerns the extent of the employee's rights, not the existence of the basic right itself. For example, there is disagreement about whether hiring undercover investigators to discover drug users at the work site is an unwarranted intrusion on the employee's rights, but there is no debate about the principle of exemption from unwarranted intrusion.

### The Employee's Obligations

In addition to enjoying rights, an employee entails obligations to various stakeholders. These obligations can form a clear and reasonable framework for discussing the ethics of technical communication. The following discussion outlines three sets of obligations—to the employer, the public, and the environment—that often conflict and thereby give rise to ethical dilemmas.

### The Employee's Obligations to the Employer

The employee's primary obligation is to further the aims of the employer and to refrain from any activities that run counter to those aims. Perhaps the most basic obligation of the employee to the employer is what is called *competence and diligence*. Competence refers to the employee's skills; he or she should have the training and experience to do the job adequately. Diligence simply means hard work; the employee is obligated to devote his or her energies to the task.

A second set of obligations is called *honesty and candor*. The employee should not steal from the employer. Stealing involves such practices as embezzlement, but also includes more common occurrences such as "borrowing" office supplies and padding expense accounts. Candor means truthfulness; the employee should report problems to the employer that will or might affect the quality or safety of the product or service that the organization provides. If, for instance, a scientist has learned that a chemical that her company is considering manufacturing and selling might be harmful to the drinking water supply, she is obligated to inform her supervisor of this fact, even though the news might displease the supervisor.

Another kind of problem involving honesty and candor concerns what Sigma Xi, the Scientific Research Society, calls trimming, cooking, and forging [23, p. 11]. Here are Sigma Xi's definitions of the three terms:

- trimming: the smoothing of irregularities to make the data look extremely accurate and precise
- cooking: retaining only those results that fit the theory and discarding others
- forging: inventing some or all of the research data that are reported, and even reporting experiments to obtain those data that were never performed.

In carrying out research, an employee might feel some pressure to report positive, statistically significant findings and must resist the temptation to indulge in these forms of dishonesty.

A third obligation of the employee to the employer is *confidentiality*. The employee should not divulge company business outside of the company. If a competitor knew that the company is planning to introduce a new product, the competitor perhaps could use that knowledge to hurt the company by introducing its own version of that product, thereby robbing the company of its competitive edge in the marketplace. Many other kinds of privileged information, such as internal problems of quality control, personnel matters, relocation or expansion plans, and financial restructuring, also could be used against the company. A well-known problem of confidentiality involves insider information; an employee knows about a development that is going to increase the value of the company's stock, and secretly buys the stock before the information is made public, thus reaping an unfair profit.

A fourth obligation of the employee to the employer is loyalty. The employee should act in the employer's interest, not in his or her own. Therefore, it is unethical for the employee to invest heavily in a competitor's stock, because that could jeopardize his or her objectivity and judgment. For the same reason, it is unethical to accept bribes or kickbacks. It is unethical for the employee to devote considerable time to moonlighting, because the outside job can create a conflict of interest and because the heavy work load can make the employee a less productive worker in the primary position.

*Employee's Obligations to the Public*

Every organization that produces products or services has certain obligations to treat its customers fairly. An employee representing an organization—and especially an employee communicating technical information—bears a portion of the responsibility for ensuring that the organization fulfill its obligations to the public. As a result, unfortunately, the technical communicator will frequently confront a conflict of goals: his or her obligation to the employer suggests one course of action, whereas his or her obligation to the public suggests another. The technical communicator can face a serious dilemma when the employer wants the public to view the product only in the most favorable light, but the public wants to know its shortcomings and limitations.

In general, an organization is acting ethically if the product or service it is selling is both safe and effective. It must not injure or harm the consumer, and it must fulfill the function it was intended to fulfill when the consumer purchased it. However, these common-sense principles provide little guidance for dealing with the complicated ethical problems that arise routinely.

Product-related accidents are commonplace. The major cause of death of people under the age of thirty, for example, is automobile accidents. A quarter of a million Americans are injured each year by power tools, and several hundred die from their injuries. Almost two million a year are injured in home construction projects; more than a thousand die. The financial losses from injuries total many billions of dollars each year. Even more commonplace, of course, are product failures or inadequacies: the items are difficult to assemble or operate, they don't do what they are supposed to do, they break down, or they require more expensive maintenance than indicated in the product information.

Who is responsible: the company that produces the product or service, or the consumer who purchases it? In individual cases of injury or product failure, it is sometimes easy enough to fix blame. If a person operates a chain saw without having read the safety warnings and without having received any instruction in how to use it, most people would say that he or she is to blame for any injuries caused by the normal operation of the saw. On the other hand, if the manufacturer knew that the chain on the saw is liable to break when used under certain circumstances but failed to remedy this problem or warn the consumer, then the company would seem to bear the responsibility for any resulting accidents.

However, cases such as these do not provide a rational theory that can help people understand how to act ethically in fulfilling their obligations to the public. According to Beauchamp and Bowie there are three main theories that describe the obligations to the public: the contract theory, the due-care theory, and the strict-liability theory [24].

The contract theory holds that when a person buys a product or service, he or she is entering into a contract with the manufacturer. The manufacturer (and by implication the employee representing the manufacturer) has four main obligations:

1. To make sure the product or service complies with the contract in several respects: it should do what it is advertised as being able to do, it should operate a certain period of time before needing service or maintenance, and it should be at least as safe as the product information explicitly states and the advertising implicitly suggests;
2. To disclose all pertinent information about the product or service, so the potential consumer can make an informed decision on whether to purchase
3. To avoid misrepresenting the product or service; and
4. To avoid coercion. The most commonly cited instance of coercion involves an unethical funeral director who takes advantage of a consumer's emotional state to sell the person a more expensive product than he or she would ordinarily contract to buy.

Critics of the contract theory argue that the typical consumer is in no position to understand the product as well as the manufacturer does, and that therefore the contract is invalid because of the consumer's ignorance.

The due-care theory places somewhat more responsibility on the manufacturer. This theory holds that the manufacturer knows more about the product or service than the consumer does, and therefore has a greater responsibility than the consumer does to make sure the product or service complies with all its claims and is safe. Therefore, in designing, manufacturing, testing, and communicating about a product, the manufacturer has to make every effort to ensure that the product will be safe and effective when used according to the instructions. However, the manufacturer is not liable when something goes wrong that it could not foresee or prevent. Critics of the due-care theory argue that because it is almost impossible to determine whether the manufacturer has in fact exercised due care, the theory offers little of practical value.

The strict-liability theory goes one step further than the due-care theory. Under the strict-liability theory, the manufacturer is at fault when injury or harm occurs from any product defect, even if the manufacturer exercised due care and could not possibly have predicted the failure. This theory is based on the premise that the only way to assess blame is to hold the manufacturer guilty and thereby force it to assume all liability costs. This way, the manufacturer will build these costs into the price of the product, and society will thus be able to afford the product. Critics of the strict-liability theory hold that, although it might be practical, it is unfair, because no organization should be held liable for something that is not its fault.

Regardless of which of these three theories the company subscribes to, the technical communicator fulfills a critical function as the link between the product and the consumer. If the technical communicator is lucky enough to work in an organization with an enlightened approach to communicating fully and fairly, he or she will rarely experience a conflict of goals. However, most technical communicators will frequently feel that they are not telling the consumer the whole truth about the product. The line between legitimate marketing and product information that concentrates on a product's strengths on the one hand, and illegitimate withholding of crucial negative information on the other, is often quite fine. Still, if the technical communicator feels that he or she is being asked to bend the truth or withhold important information pertaining to a product's general quality, useful life, or safety, he or she should go to management and try to resolve the dilemma.

### Employee's Obligations to the Environment

Perhaps the most important lesson we have learned in the last decade is that our natural resources are limited and that we are polluting and depleting them at an unacceptably high rate. The overreliance on fossil fuels not only deprives future generations of their use, but also causes terrible pollution problems that many scientists believe are irreversible, such as global warming. Everyone—government, business, and individual—must work to preserve the fragile ecosystem, to ensure the survival not only of our own species but also of the other species with which we share the planet.

But what does this have to do with technical communication? Technical communicators, in their daily work, do not cause pollution or deplete the environment in any unique or extraordinary way. Yet because of the nature of their work, they are often aware of the environmental impact of a policy or action that their organization has implemented or taken. For example, a writer or graphic artist who is working on a proposal that is to be submitted to the federal government might help create the environmental impact statement. What the technical communicator says or does not say—or insists on saying—can often have an enormous effect on the environment. For instance, if the communicator knows that a scientist at the company has discovered that a planned construction project would have a serious negative effect on an animal species living in the affected area, he or she should feel obligated to discuss that impact, even if management wishes to omit it or downplay it.

Situations involving possible or definite negative effects on the environment are sometimes quite complicated ethically. One frequent problem is lack of solid information, making it impossible to assess accurately the potential for environmental damage. For instance, the most informed scientific opinion might be divided on whether harvesting lumber in a particular area will seriously affect a rare species of owl, yet everyone agrees that the harvesting will provide a necessary boost to the economy of the region. Given that there is a known correlation between the economic depression of a region and its rates of alcoholism, domestic abuse, and mental illness, what should the technical communicator do? Try to tell the truth, describing as accurately as possible only what is known and what is unknown.

Technical communicators should be sensitive to environmental ethics and treat every occurrence or potential occurrence of environmental damage seriously. They should make sure they supervisors are aware of the situation and try to work with them to reduce the actual or potential damage to the extent possible. The conflict that all people face is, of course, that protecting the environment is expensive. Profitable projects are delayed, made more costly, or prevented entirely; cleaner fuels are more expensive than dirty ones, and disposing of hazardous waste properly is much more costly (in the short run) than merely dumping it. In a business environment dominated by a desire to cut bottom-line costs, the temptation to cut corners on environmental protection will be strong.

## ANALYZING COMMON DILEMMAS FACED BY A TECHNICAL COMMUNICATOR

An employee will likely confront both simple and complicated ethical dilemmas from the first day on the job. Should the employee use company pens and stationery for personal business? This is obviously a simple ethical question. But if an employee accepts an attractive job with one of her previous employer's competitors, only to discover that the new company wants her to divulge information about her previous employer's products, then she is facing a much more complicated ethical dilemma.

Most of the difficult ethical dilemmas involve a conflict between two competing principles. For example, in the question of the new company's wanting the employee to provide secret information about a competitor, the dilemma is between loyalty to her current employer and the prohibition against stealing (stealing information, in this case). The following discussion suggests some of the ways that people have tried to address several of the more common ethical dilemmas faced by professional people in their roles as communicators:

plagiarism, trade secrets, fair use of language and visuals in product information and advertising, and whistleblowing.

## Plagiarism

Although most people know that it is unethical, plagiarism is in fact a complicated issue, especially in modern technical communications. In many organizations, authorship of internal documents such as memos and most reports is regarded casually; that is, an employee who is asked to update an internal procedures manual will be expected to use any material from the existing manual, even though he or she does not know and perhaps even could not determine the original author.

For documents that are to be published, such as external manuals or journal articles, authorship is treated much more carefully. Because most documents of this kind are produced collaboratively—with several persons contributing text, another one doing the graphics, a third person reviewing for technical accuracy, and a fourth reviewing for legal concern—it is often quite difficult to determine who "wrote" it.

The best way to approach the question of authorship is to do so openly, by discussing it with all persons who contributed to the document. Some persons might want and deserve to be listed as authors; others might more appropriately be credited in an acknowledgement section. To prevent charges of plagiarism, the wisest course is to be very conservative: if there is any question about whether to cite a source, it should be cited.

## Trade Secrets

What is a trade secret? According to the law, a trade secret is "any confidential formula, pattern, device, or compilation of information that is employed in a business and that gives the business the opportunity to gain a competitive advantage over those who do not know or use it" [24, p. 264]. For example, the formula for the syrup that Coca-Cola uses to make its soft drink is a trade secret, even though the individual ingredients of the syrup are commonly available. According to the law, if a scientist working for Coca-Cola were to quit and join a competing company, he or she would not be permitted to divulge the formula to the new employer. Coca-Cola owns that information.

The problem is that in most cases it is very difficult to define what constitutes a trade secret. If, for instance, over the course of a nine-year career with a particular company a systems analyst devises a unique approach to structured programming that his company employs in many of its products, who owns that approach: the analyst or the company? That approach gives the company a competitive advantage, and so it would seem to be a trade secret. Yet the systems analyst devised that approach, and one of his fundamental rights is the ability to work for whatever employer he chooses.

An additional complication is that many disclosures of information are unintentional. If an engineer works at a particular company for a number of years, the line between what information he brings to the company and what information he gains from working there blurs. Through his work experience he develops a way of thinking, a way of approaching problems. It might be impossible for him to state what is his company's legitimate trade secret and what is his own thinking.

Many companies are trying to cut down the number of problems involving trade secrets by developing new management practices, such as entering into contracts with certain key individuals explicitly forbidding the use of some trade secrets, restricting the number of persons who are exposed to the crucial information, and preventing employees from engaging in outside consulting. Other companies use positive incentives, such as generous salaries and post-retirement consulting contracts, to keep their employees from wanting to work for their competitors.

Employees today tend to change jobs much more frequently than they did several decades ago, and if they are to retain this mobility, companies will find it increasingly difficult to keep information secret. The law does not provide clear, precise guidelines on the question of what constitutes a trade secret and therefore the question will likely remain an ethical dilemma that the individual must confront and try to resolve.

## The Fair Use of Language and Visuals

Employees who write or create visual information will often face ethical dilemmas as they try to communicate fairly and effectively. In writing a proposal, for example, an employee might feel under some pressure to exaggerate claims about the company's expertise or experience or to minimize or even ignore disadvantages of the proposed plan. As Louis Perica has pointed out it is a relatively simple matter to alter drawings and photographs to unfairly influence the reader [25]. Now that photographs are routinely digitized, the potential for abuse is increased.

Ethical problems are particularly common in product information, from descriptions in sales catalogs to specification sheets and operating instructions and manuals. As Powledge points out, in most cases the cause of the ethical dilemma is that in creating product information, the writer is doing two things at once: describing and advertising the product [26]. These two functions are not only different, they are often in conflict. (Ironically, most advertisers reject this distinction; they maintain they are merely providing product information. Yet it is hard to find much objective information in most ads.)

Employees report that sometimes they are asked to lie or mislead the reader. Lying—providing false information—is obviously unethical in these situations.

If a company's own tests of the disk drive it sells show that the mean time between failures (MTBF) is 1,000 hours, but the competitor's figure is 1,500 hours, the writer at the company might be pressured by a supervisor to simply lie, to say that the MTBF is 1,750 hours.

Providing misleading information is a little more complicated, but it amounts to the same thing ethically. A misleading statement or visual, while perhaps not actually being a lie, enables or even encourages the reader to believe false information. For instance, a product-information sheet for a computer system is misleading if the accompanying photograph of the unit includes a modem but in fact the modem is sold separately and is not part of the purchase price listed on the sheet.

Misleading information can be communicated in a number of other ways. The writer can try to scare the reader by falsely suggesting that a product—or a particular brand of the product—is necessary. For example, a flashlight manufacturer is being misleading in suggesting that only its own brand of batteries will power the flashlight. Another technique is to ignore the negative features of a product. For instance, if an information sheet for a portable compact disc player suggests that it can be used by joggers, without mentioning that it will probably skip from the bouncing, that is misleading. A third technique of providing misleading information is the use of legalistic syntax that doesn't mean what it seems to mean. For instance, it is unethical to write, "The 300X was designed to operate in extreme temperatures, from -40 degrees to 120 degrees Fahrenheit," if in fact the unit cannot reliably operate in those temperatures. The fact that the statement might actually be accurate—the unit was *designed* to operate in these temperatures—doesn't make it any less misleading.

William James Buchholz describes three other characteristics of writing that can mislead readers, regardless of the writer's intentions: 1) abstractions and generalities, 2) jargon ("user friendly"), and 3) euphemism ("market research" instead of a "few phone calls") [27].

## Whistleblowing

Whistleblowing occurs when an employee goes public with information about an unethical act or practice within his or her company. For example, an engineer is blowing the whistle when she tells a government regulatory agency or a newspaper that quality-control tests on one of the products her company sells have been faked.

The dilemma that an employee faces in the case of whistleblowing is between loyalty to the employer, on the one hand, and to his or her own standards of ethical behavior, on the other. Some people believe that an employee owes complete loyalty to the employer. The former president of General Motors, James M. Roche, for example, has written,

> Some of the enemies of business now encourage an employee to be *disloyal* to the enterprise. They want to create disharmony, and pry into the proprietary interests of the business. However this is labeled—industrial espionage, whistle blowing, or professional responsibility—it is another tactic for spreading disunity and creating conflict [24, p. 262].

However, many would still argue that no employee owes an employer complete loyalty. For example, most reasonable people would agree that workers should not be asked to steal or lie or take actions that could physically harm others. Therefore, the question is: where does loyalty to the employer end, and at what point does the employee have a right to take action? And what is the proper procedure for dealing with a serious ethical problem before resorting to whistleblowing? Ethicist Manuel Velasquez outlines four questions that an employee should consider carefully before taking any action:

1. *Does the employee understand the situation fully and are his or her facts accurate?* Sometimes employee's grievances are the result of incomplete or erroneous information.
2. *What exactly is the ethical problem involved?* The employee should be able to explain what is unethical about the practice and state just who is being harmed by it.
3. *Is the ethical problem serious or trivial?* The more serious the ethical problem, the more the employee is justified or even obligated to take some action.
4. *Would the employee be more effective in stopping the unethical practice if he or she worked privately within the organization or blew the whistle?* What are the implications—for society, for the company, and for the employee—of each of the two paths? [22, p. 381].

These questions suggest that whistleblowing is a very serious act that should not be performed casually. The employee should make every effort to solve the problem internally before going public. A number of organizations have instituted procedures to try to encourage employees to bring ethical questions to management. Among the more common tactics are the use of anonymous questionnaires and the establishment of a position of ombudsman, whose job is to bring ethical grievances to management's attention. If the ombudsman feels that management has not responded satisfactorily to the situation, he or she is empowered to report the information freely.

However, most companies still have no formal procedures for handling serious questions about ethical matters. For this reason, whistleblowing remains a risky action for the employee to take. Although the federal government and about half the states have laws intended to protect whistleblowers, these laws are not highly effective. It is simply too easy for the organization to penalize the whistleblower—subtly or unsubtly—through negative performance appraisals, transfers to undesirable locations, or other forms of isolation within the company. For this reason, many people feel that an employee who has unsuccessfully tried every method of alerting management to a serious ethical problem would be wise to simply resign rather than face the professional risks of whistleblowing. Of course, resigning quietly is much less likely to force the organization to remedy the situation. As many ethicists say, doing the ethical thing is not always in a person's best interest professionally.

## THE ROLE OF THE CODE OF CONDUCT

The kinds and numbers of problems that lead to whistleblowing will probably increase along with the public's heightened interest and understanding of the role of ethical behavior. To try to reduce the incidence of unethical behavior, many businesses and professional organizations have written and distributed professional codes of conduct. As of 1985, three quarters of the 1,500 largest American corporations had written them, as had virtually all professional organizations [3, p. 91].

Codes of conduct vary greatly from organization to organization, but all include statements of ethical principles that the employees of the company or members of the organization should follow. Some are brief and general, offering only guidelines for proper behavior. The Code of Communicators written by the Society for Technical Communication, for example, is less than 200 words long and consists of statements such as "Because I recognize that the quality of my services directly affects how well ideas are understood, I am committed to excellence in performance and the highest standards of ethical behavior."

Other organizations' codes go into great detail in describing proper and improper behavior and actually stipulate penalties for violating the principles. These codes could be thought of more as sets of rules than as general guidelines. The American Society of Mechanical Engineers' code, for instance, specifies procedures the Society is to follow in cases of complaints about ethical violations. These procedures are quite specific about the proper functioning of its Professional Affairs and Ethics Committee, even indicating, for instance, that the complaint must be acknowledged "by Certified Mail" [28, p. 193].

Do codes of conduct really work in encouraging ethical behavior? This is not an easy question to answer, of course, because there are no statistics on how many people didn't act unethically because they were inspired or frightened into acting ethically by a code of conduct. However, many ethicists and officers in professional societies are skeptical. Because codes have to be flexible enough to cover a wide variety of situations, they tend to be so vague that it is virtually impossible to say that a person in fact violated one of their principles.

A study conducted by the American Association for the Advancement of Science in 1980 found that whereas most professional organizations in the sciences and engineering have codes, and many of these codes do contain provisions for hearing cases of unethical conduct, relatively few such cases have ever been brought before the organizations [28]. Many societies reported, for example, that only three or four allegations were ever lodged against individuals, and most of the societies have never taken disciplinary action—censure or expulsion—against any of their members for ethical violations.

If codes of conduct are not often applied in any systematic way, do they have any real value? Ethicist Jack N. Behrman points out that perhaps the greatest value of a code comes from the mere process of writing it, because it forces an organization to clarify its own position on ethical behavior [29, p. 156]. Of course, the act of distributing the code might also have the effect of fostering among employees or members of the organization an increased awareness of ethical issues, which in itself can only be positive.

Critics of codes of conduct point out that almost no organization is willing to support someone who brings a charge against someone else. Because an accuser within an organization is likely to face oven or covert punishment, as mentioned earlier in the discussion of whistleblowing, self interest compels many people to remain silent, regardless of the high-sounding statements in the company's code of conduct. Few professional organizations have ever come to the financial aid of an accuser who has lost a job because of a justified allegation.

For this reason, codes of conduct are sometimes thought of as primarily public relations tools intended to persuade not only employees and organization members, but also the general public and the government. With these codes of conduct, the organization polices its own members so that the federal government will not impose its own regulations.

Creating a code of conduct, therefore, can be a cynical exercise in manipulating employees and the public, or it can a constructive process for improving the employees' understanding of ethical issues and their adherence to high ethical standards. Ultimately, however, the code of conduct is less important than the ethical atmosphere in the organization. If management acts ethically and hires and promotes people who also act ethically, the corporate culture at that organization will perpetuate a high degree of awareness about ethical issues. If management does not embody high ethical standards, the most thorough code of conduct will have no real effect.

## USING THE CASE METHOD

Examining the ethics of technical communication from the perspective of literature of business and professional ethics will not, of course, provide a simple method for resolving the many and serious ethical dilemmas that technical communicators routinely face on the job. However, by devoting instructional time to the issue of ethics, we demonstrate to our students the value we place on ethical conduct. And by establishing the basic framework of the employee's obligations to the employer, the public, and the environment, we place the common ethical issues, such as plagiarism and the fair use of language and visuals, within a rational context, and thereby, clarify the writer's choices.

One of the most popular ways to teach ethical thinking is through the use of the case method, either real cases (such as the BART whistleblowing case or the Ford Pinto case) or simulated. In the most effective cases, the protagonist faces a clear conflict of goals; the students, therefore, can come to different conclusions about the proper course of action.

Cases appear frequently in professional journals. Volumes of cases include those by Baum and Flores 1980 [30] (engineering), Beauchamp 1983 [31] (business), Donaldson 1983 [32] (business), Goodpaster 1984 [33] (management), Veatch 1977 [34] (medicine), and Velasquez 1982 [22] (business).

Following is a sample case that draws upon a number of the theoretical issues developed in this essay. The protagonist in the case is a technical communicator who must confront a set of facts that are a lot less conclusive than she would like. Any decision she will make will have both positive and negative outcomes.

### The Diversified Construction Materials Case

The town of Acton, Ohio (population 6,500), like many other small communities in the Rust Belt, has suffered economically during the last decade. Much of its infrastructure is old and in need of repair, but the town has a shrinking tax base. Young people routinely leave after high school in search of better employment opportunities.

The main employer in Acton is Diversified Construction Materials, which employs over 1,000 people from Acton and surrounding communities. Like Acton, Diversified has known better times. Its products are known for their high quality, but foreign manufacturers and domestic manufacturers who have moved their production facilities to third-world countries are undercutting Diversified's prices and gaining market share.

However, the Research and Development Department at Diversified has just formulated a new type of blown insulation that the company thinks will perform as well as fiberglass but beat the price. This new substance promises to be a major part of Diversified's highly-regarded product line in insulation. A number of retailers have placed large orders of the new insulation based on Diversified's exhibits at trade shows and some preliminary advertisements in industrial catalogs.

As the head technical writer at Diversified, Susan Taggert has been assigned the task of overseeing the creation of all the product information for the insulation. As she normally does in such cases, she gathers all the documentation from R&D and any other relevant materials, which she will study before mapping out a strategy.

About one week into the project, Taggert discovers from laboratory notebooks that three of the seven technicians working on the project experienced abnormally high rates of absence during their four months working on the insulation. One of the three technicians requested to be transferred from the project at the end of the first month. His request was granted.

Calling Diversified's Personnel Department, Taggert learns that all three of the technicians complained of the same condition, bronchial irritation, to varying degrees of severity, but that the irritation ceased two to three days after the last exposure to the insulation. Apparently some compound in the insulation, which the company physician could not identify, affected some of the technicians who worked closely on it.

Taggert goes to the Vice President of Operations, Harry Mondale, who is in charge of the introduction of all new products. Taggert presents her information to Mondale and suggests that the company find out what is causing the bronchial irritation before it ships any of the product. Although the irritation does not appear to be serious, there are no data on the potential effect of long-term exposure to the insulation when used in houses or offices. Mondale points out Diversified's tight deadline; delivery is scheduled in less than two weeks. Determining the cause of the irritation could take weeks or months and cost many thousands of dollars. Moreover, Mondale points out, the Occupational Safety and Health Administration (OSHA) has already approved Diversified's application to manufacture and sell the product.

Taggert points out the financial risks involved in selling a product that poses a health risk. Mondale responds that that is a risk the company will have to take. The company has staked its reputation—and its third-quarter profits—on the insulation. He directs Taggert to proceed with the product literature as quickly as possible and not to spend any more time worrying about the health hazard.

What should Susan Taggert do?

This case describes a typical conflict of goals. Susan Taggert wishes to demonstrate loyalty to employer, who wants to go ahead with the sale of the product, yet she feels an obligation to a number of other affected stakeholders. In response to her boss' claim that the welfare of the company (and, by implication, the community) depends on selling the insulation, she argues that the long-term interests of the company could be seriously jeopardized if the product turns out to be unsafe. In addition, she fears that an unsafe product could injure the company employees who work in it, the contractors who install it, and the people who live or work in buildings that contain it. Because this dilemma involves serious safety risks to innocent people, Taggert is ethically obligated to pursue the matter. She might consider writing a formal letter to her boss, Mondale, with a copy to his supervisor, carefully explaining her argument: that the product should be withheld from the market until more extensive tests can be conducted that will determine whether the product is safe to manufacture and distribute. If nobody in the company will agree to take this action, she should consider blowing the whistle.

## REFERENCES

1. H. N. Shirk (ed.), *Bulletin of the Association for Business Communication, 53*:3, 1990.
2. A. Tibbetts, (ed.), *Journal of Business Communication, 22*:1, 1985.
3. J. Brockmann and F. Rook (eds.), *Technical Communication and Ethics,* Society for Technical Communication, Washington, D.C., 1989.
4. S. Doheny-Farina (ed.), *IEEE Transactions on Professional Communication, 30*:1, 1987.
5. C. H. Sides (ed.), *Technical and Business Communication: Bibliographic Essays for Teachers and Corporate Trainers*, National Council of Teachers of English, and Society for Technical Communication, Urbana, Illinois, 1989.
6. R. Rentz and M. B. Debs, Language and Corporate Values: Teaching Ethics in Business Writing Courses, *Journal of Business Communication, 24*:3, pp. 37-48, 1987.
7. F. B. Krohn, A General Semantics Approach to Teaching Business Ethics, *Journal of Business Communication, 22*:3, pp. 59-66, 1985.
8. M. Kremers, Teaching Ethical Thinking in a Technical Writing Course, *IEEE Transactions on Professional Communication, 32*:2, pp. 58-61, 1989.
9. H. Sachs, Ethics and the Technical Communicator, in *Technical Communications and Ethics,* R. J. Brockmann and F. Rook (eds.), Society for Technical Communication, Washington, D.C., pp. 7-10, 1989.
10. C. Yee, Technical and Ethical Professional Preparation for Technical Communication Students, *IEEE Transactions on Professional Communication, 31*:4, pp. 191-198, 1988.
11. R. L Johannesen, Teaching Ethical Standards for Discourse, *Journal of Education, 162*:2, pp. 5-20, 1980.
12. M. R. Wicclair and D. K. Farkas, Ethical Reasoning in Technical Communication: A Practical Framework, in *Technical Communication and Ethics,* R. J. Brockmann and F. Rook (eds.), Society for Technical Communication, Washington, D.C., pp. 21-25, 1989.
13. S. Golen, C. Powers, and M. A. Titkemeyer, How to Teach Ethics in a Basic Business Communication Class—Committee Report of the 1983 Teaching Methodology and Concepts Committee, Subcommittee 1, *Journal of Business Communication, 22*:1, pp. 75-83, 1985.
14. J. C. Condon, Jr., *Interpersonal Communication*, Macmillan, New York, 1977.
15. P. Rubens, Reinventing the Wheel? Ethics for Technical Communicators, in *Technical Communication and Ethics,* R. J. Brockmann and F. Rook (eds.), Society for Technical Communication, Washington, D.C., pp. 15-20, 1989.

16. B. A. Spencer and C. M. Lehman, Analyzing Ethical Issues: Essential Ingredient in the Business Communication Course of the 1990s, *Journal of the Association for Business Communication*, 53:3, pp. 7-16, 1990.
17. G. Kennedy, *The Art of Persuasion in the Roman World (300 BC - AD 300)*, Princeton University Press, Princeton, New Jersey, 1972.
18. G. Kennedy, *The Art of Persuasion in Greece*, Princeton University Press, Princeton, New Jersey, 1963.
19. B. Vickers, *In Defence of Rhetoric*, Clarendon Press, Oxford, 1988.
20. G. E. Yoos, A Revision of the Concept of Ethical Appeal, *Philosophy and Rhetoric*, *12*:1, pp. 41-58, 1979.
21. R. Baumhart, *An Honest Profit: What Businessmen Say About Ethics in Business*, Holt, Rinehart, and Winston, New York, pp. 11-12, 1968.
22. M. G. Velasquez, *Business Ethics: Concepts and Cases*, Prentice-Hall, Englewood Cliffs, New Jersey, 1982.
23. *Honor in Science*, Sigma Xi, The Scientific Research Society, New Haven, Connecticut, 1986.
24. T. L Beauchamp and N. E. Bowie, *Ethical Theory and Business*, (3rd Edition), Prentice-Hall, Englewood Cliffs, New Jersey, 1988.
25. L. Perica, Honesty in Technical Communication, *Technical Communication, 15*:1, pp. 2-6, 1972.
26. T. M. Powledge, Morals and Medical Writing, *Medical Communication, 8*:1, pp. 1-10, 1980.
27. W. J. Buchholz, Deciphering Professional Codes of Ethics, *IEEE Transactions on Professional Communication, 32*:2, pp. 62-68, 1989.
28. R. Chalk, *AAAS Professional Ethics Project: Professional Ethics Activities in the Scientific and Engineering Societies*, American Association for the Advancement of Science, Washington, D.C., 1980.
29. J. N. Behrman, *Essays on Ethics in Business and the Professions*, Prentice-Hall, Englewood Cliffs, New Jersey, 1988.
30. R. J. Baum and A. Flores (eds.), *Ethical Problems in Engineering*, (2nd Edition, 2 vols.), Center for the Study of the Human Dimensions of Science and Technology, Rensselaer Polytechnic Institute, Troy, New York, 1980.
31. T. L. Beauchamp, *Case Studies in Business, Society, and Ethics*, Prentice-Hall, Englewood Cliffs, New Jersey, 1983.
32. T. Donaldson, *Case Studies in Business Ethics*, Prentice-Hall, Englewood Cliffs, New Jersey, 1983.
33. K. E. Goodpaster, *Ethics in Management*, Harvard Business School, Boston, 1984.
34. R. M. Veatch, *Case Studies in Medical Ethics*, Harvard University Press, Cambridge, Massachusetts, 1977.

# POLITICAL-ETHICAL IMPLICATIONS OF DEFINING TECHNICAL COMMUNICATION AS A PRACTICE

## Dale L. Sullivan

Let me present one possible version of the history of teaching writing in the last century and a half. When the tradition of classical rhetoric was restricted to composition in the nineteenth century, teachers of writing found themselves teaching service courses, usually defined as skills courses. Furthermore, having lost touch with the classical tradition, they began to teach writing particularly suited to current needs and, by extension, to teach thought forms that imitate modern consciousness—a form of consciousness largely molded by forms of production, or technology. As Richard Ohmann says, much modern composition instruction reflects this technological consciousness: it casts the writing process in terms of problem solving, stresses objectivity and thereby denies a writer's social responsibilities, distances the interaction between writer and reader, deals with abstract issues, and denies politics [1]. As a result, teachers of writing indoctrinate students, turning them into the sorts of people who will fill the slots available in our technological society.

If this story is a suggestive account of rhetoric's metamorphosis into composition, it is even more interesting applied to rhetoric's transformation into technical communication. Rhetoric has always aimed at teaching professional discourse—particularly the discourse of the assembly, the court, and later the pulpit—and so it is possible to see technical communication as a direct descendant of rhetoric, even more in tune with its aims than is composition. However, though technical communication shares classical rhetoric's orientation toward the professions, those of us who teach technical communication don't often think of ourselves as carrying on the rhetorical tradition. Indeed, it is rather hard to do so, since we teach thought forms and discourse forms demanded by the workplace, and we often find ourselves representing the military-industrial complex instead of the humanistic tradition. As John Mitchell puts it, we "indoctrinate our students in the forms appropriate to their employers," for "the students know they must dance with the guy that brung them, and they elect our courses to learn his dance steps" [2, p. 5]. In fact, the social contract that legitimizes the teaching of technical writing seems to insist that we adopt the technological mindset. For example, J. C. Mathes, Dwight Stevenson, and Peter Klaver warn engineer teachers that it is dangerous to let people trained in classical rhetoric and literature teach technical writing, because in so doing they "risk having their students taught principles that are in conflict with engineering principles" [3, p. 332].

Another way of looking at this situation is to say, as Michael Halloran has, that we perpetuate the "ethos" of the technological society, primarily by viewing the rhetorical art as "a set of technical skills practiced by specialists" [4, p. 221]. These skills are forms of technology—they are *techne*, to use Aristotle's term— and, as such, their products can be separated from the maker and marketed, relieving the writer of responsibility [4, p. 227]. As teachers of composition or technical writing, we sometimes find this project something we can live with, but there exists a fairly long tradition of reaction against it. Nan Johnson, for instance, documents the attempts of three nineteenth-century rhetoricians—Theremin, Day, and Hope—to stand against the reduction of rhetorical education to the teaching of specialized skills; and Robert Connors describes the continuing battle over the issue of humanism versus vocationalism in technical communication, a battle that has apparently been part of the profession from its inception early in this century [5, 6].

## TECHNICAL GENRES AND POLITICAL-ETHICAL STASIS

In the past ten to fifteen years, several articles have contributed to this ongoing debate, but it is not my purpose to catalog them. Instead, I would like to focus on three essays by Carolyn Miller that, I think, lead to a point of political and ethical stasis for those of us who teach technical communication. In the first article, *A Humanistic Rationale for Technical Writing* (1979), Miller argues that traditional technical writing instruction is based on the "windowpane" theory of language, a theory that frames technical and scientific writing as "just a series of maneuvers for staying out of the way" [7, p. 613]. If we discard this antiquated view of language, Miller says, teaching technical writing can be more than teaching a set of skills; instead, it can be a "kind of enculturation" that helps students understand how to belong to a community [7, p. 617]. This conception of teaching technical writing has the virtue of fitting nicely with students' definitions of the course; that is, it is a course that gives them passage "in" to a certain group [8, p. 23]. More to the point, it offers an important advance over the skills-based approach to teaching technical writing. Nevertheless, it leaves unanswered a crucial question: what are we enculturating our students into?

To answer this question, we must take up the issue of genre, for genres are schemas of response considered appropriate by a discourse community [9]. Clearly, these schemas are not value neutral; when students learn them, they learn what may be said about possible subjects on particular occasions [10, p. 165]. In other words, genres change the way we think by defining rhetorical situations—hat the Greeks called *kairos,* or the opportune moment [11]. Thus, as Patricia Bizzell says, it is "difficult to maintain the position that discourse conventions can be employed in a detached, instrumental way" [12]. Unfortunately, genres in technical discourse seem to preclude the opportunity for citizens to speak simply as citizens on the issues of technology in any meaningful way. So one way we enculturate students is by teaching them the genres of technical discourse, though the concept of genre is often reduced to the notion of form. As Connors has shown, teaching technical forms has been a long-standing tradition among technical writing teachers [6, p. 338], a tradition still followed by many today.

At this point, Miller's *Genre as Social Action* (1984) becomes pertinent [10]. Miller's own definition of genre as "social action" leads her to deny that certain technical forms, specifically environmental impact statements, are genres because they preclude social action [10, p. 164]. That is, because these forms attempt to incorporate the interests of several factions, the writer becomes mechanized and, in Burkean terms, produces motion rather than action. I have little doubt that careful study of other standardized technical forms, whether governmental or industrial, would suggest that Miller's observations obtain widely. My own conclusion, therefore, is that teaching standardized formats and forms means teaching the technological mindset, and, thus, enculturating students into the military-industrial complex. This conclusion further suggests that we implicitly accept present restrictions on public discourse about technology and fail to give students power to engage in social action.

The third article that bears on this issue is Miller's *What's Practical about Technical Writing?* (1989) [13]. In this paper, Miller suggests that we define technical writing as *praxis* rather than as *techne, praxis* being the Greek word for social action and *techne* for an art of making. This move allows her to recommend that we question present practices, ask our students to do the same, and encourage them to take socially responsible action. As this recommendation suggests, Miller's three articles show an evolution in her thought, always in a consistent direction, but they also lead to a point of stasis: if we enculturate students in the technical writing classroom, at least in part by teaching technical genres that reinforce the dominance of the technological system, how can we then call them to responsible social action?

## *PRAXIS,* VIRTUE, AND SOCIAL ACTION

I wish to suggest that this conflict is only an apparent paradox and that those of us who teach the course are really placed in a situation that allows us to be powerful agents for change. But to have a class that encourages social action requires adopting Miller's suggestion that we define technical communication as a practice rather than as an art or skill. As she points out, Aristotle makes a distinction between the ability to produce products, a technical skill that he calls *techne,* and the ability to take social action, or *praxis* [14, sect. 6.4]. Further, the ability to take social action involves the virtue of practical wisdom or prudence (*phronesis*), a virtue defined as the ability to reason about ends rather than means. Phronesis enables a person to deliberate about the good rather than the expedient and, as such, to act in the political sphere rather than in the sphere of work [14, sect. 6.5]. As Barbara Warnick says, techne is "a habit of producing," whereas phronesis is concerned with the uses to which products are put [15, pp. 304-305]. Taking this distinction into account, we can define social action as action free from the economic constraints of the workplace: it is the political-ethical act of someone functioning in the citizen's role rather than in the worker's role. Unlike techne, which has an end other than itself, responsible social action constitutes eupraxia, Aristotle's word for "good action," an end in itself [14, sect. 6.5].

When rhetoric, of whatever type, is defined as a practice, it is linked with virtue. Aristotle himself does not directly link them: he defines rhetoric as an art [16, sect. 1354a.10]. However, Halloran argues that eloquence was considered a virtue by many classical rhetoricians, and Eugene Garver, Lois Self, and Oscar Brownstein all make connections between rhetoric and phronesis [4, 17-19]. Implicit in all of these studies is the definition of rhetoric as a social act or practice rather than an art, and this definition brings rhetoric out of the amoral realm of technique into the world of ethics and politics. This distinction is important, for a skill can be used for good or bad ends, but a virtue automatically embodies good ends [17, p. 69]. That is, if rhetoric is merely a skill, someone may use it to manipulate people, but if it is a virtue, then it must be used for good.

The definition of the "good," however, is problematic. In his *Ethics*, Aristotle defines it as happiness (*eudaimonia*), and happiness he defines as the virtuous activity of the soul. Furthermore, he says that virtuous activity is the ability to conform to the ideals of the society [14, sect. 1.7.7]. In other words, Aristotle's view of the good is sociological: the community defines what the good is, and the individual is good when he or she performs well the functions required by society—that is, when the person is a good citizen. Alasdair MacIntyre says much the same thing in *After Virtue*, a modernized version of Aristotelian ethics in which he approaches the subject by discussing the meaning of virtue. According to MacIntyre, the concept of virtue is embedded in at least three contexts. First, a virtue is the human quality that enables a person to engage in a practice with excellence [20, p. 191]. Second, such a quality is part of a person's complete life and character, which can be seen as an "embodiment" of a socially sanctioned narrative [20, p. 144]. hird, such socially sanctioned narratives are really roles within a larger narrative, the narrative of the culture and its tradition [20, p. 258]. Thus, we see that virtue—the good—is defined socially by a society's ideals, which, in turn, valorize roles within the society. When people fill the roles well, when they possess character traits that allow them to perform the functions of these roles with excellence, then their actions are considered virtuous.

Let us now take this depiction of ethics and apply it to our present situation. I have tentatively decided to define technical communication as a practice; therefore, I am claiming that it takes virtue to participate in technical communication. I can do this, according to Aristotelian ethics, only by agreeing that my students are developing character traits that enable them to perform their functions well. Moreover, I imply that these functions are good, that they fit in with the ideal of virtue that dominates our society.

There is no problem with this account if we are willing to accept the values embedded in the technological society, for ours is a technological society—or at least the arguments made by such social critics as Jurgen Habermas, Jaques Ellul, and William Barrett would lead us to believe that it is. But, of course, this is where we run into trouble. Many of us do not agree or identify with the values of the technological society and the military industrial complex. Instead, we identify with a variety of alternative social groups quite diverse in their plurality but all sharing at least one value: that human beings should not be subordinated to the technological imperative. As such, we want to regain the upper hand; that is, we want to make technology serve humans instead of letting technology shape our society and its values. Therefore, we can call technical communication a virtuous practice only when it is put to the service of one of these alternative humanistic visions.

But the very thought processes embodied in most modern technical genres have grown out of the technological mindset, and they continue to support the dominance of the technological society while denying people the power to take social action as citizens when they write. In effect, if we continue to teach these genres, we indoctrinate our students into a system we don't agree with; but if we stop preparing them for their roles in the technological world, then we are no longer really teaching technical communication according to the social contract that we all bought into when we agreed to teach the course.

It seems that we're back to the original point of stasis. Like David Dobrin, in *What's the Purpose of Teaching Technical Communication*, I find myself faced with a set of alternative actions I can take, though my alternatives differ from his: 1) I can get with the program, change my values, and become a representative of the technological society; 2) I can leave the profession of teaching technical writing; 3) I can become schizophrenic; or 4) I can figure out how to change my course so that it at once teaches the discourse appropriate for the technological world *and* makes students aware of the values embedded in such discourse and the dehumanizing effects of it [21]. Obviously, I think number four is the best alternative, and I would like to suggest some ways to begin teaching technical communication as a truly virtuous practice, as responsible social action.

## POLITICAL DISCOURSE IN TECHNICAL WRITING

My suggestions—to be taken as explorations of possibilities rather than as prescriptive guideline—involve altering what we teach when we teach technical communication and changing how we teach it. Altering what we teach requires redefining not only the function but also the scope of technical communication.

Certainly, we can redefine its function simply by calling it a practice, a social act, rather than an art. But we must also look at the boundaries we have drawn for technical communication, boundaries often summed up in the phrase "writing for the world of work," a phrase set in contrast to the rhetoric of leisure, as Miller points out [13, pp. 15, 18]. Classical rhetoric, though it aimed at preparing students to fill professional roles, was concerned with roles reserved for citizens, or the leisured class. We often misinterpret leisure, associating it only with elitism and forgetting that the leisured class was responsible for politics. Conversely, most of the writing done in the "world of work" was done by slaves. The ancient class distinctions implicit in classical rhetoric still carry over, even though social conditions have changed. That is, when we define technical communication as writing for the world of work, we tend to draw a boundary at the point where political discourse picks up. Within the present boundaries, technical discourse is constrained by the criteria established by industry, the division of labor within large companies, and bureaucratic procedures in government. As Susan Wells puts it, the goal of this discourse is "systematic misunderstanding and conceal-ment." She goes on to say that "the subjective responses of readers and writers are irrelevant, and the monologic voice conceals, not a dialogic relation, but the total fragmentation and dispersal of knowledge" [22, p. 256].

I think we stop short of including political discourse within the boundaries of technical communication because of the marriage between private enterprise and government bureaucracy, a system that blocks citizens from participating in effective deliberative rhetoric about the direction that technology should take [23, p. 342]. In the place of public deliberation, we have the twin motives of profit and technological advance, sacred territory in our society. There are few, if any, socially sanctioned opportunities for citizens to participate effectively in making decisions about the large issues associated with technology, or most other issues for that matter [24]. Therefore, our present way of defining technical communica-tion as the discourse appropriate for industry is equivalent to defining it as the rhetoric appropriate for slaves—those barred from making decisions about the ends, those whose decision-making authority is restricted to determining the most efficient means of obtaining predetermined ends.

If we are serious about defining technical communication as a practice, then we must expand its scope to include political discourse. To do this is to act on the ideal that all citizens, though workers, are responsible political agents; it is to act as though slavery really was abolished and not just restructured; and it is to treat the individual as a unified whole, not as a person who must divide his or her personality between the roles of the worker and the citizen. In short, expanding the scope of technical communication to include political discourse is to fight against the alienation produced by our economic and technological systems.

I am not saying we should refuse to teach the discourse appropriate for the world of work, for I think the social contract we have with our students demands that we prepare them for their future careers. But it is possible to teach this discourse from a critical perspective and to supplement it with discourse that is appropriate for social action. For example, Wells, using Habermas's ideal of communicative action, suggests that we begin by teaching the structures of "purposive-rational action"—Habermas's term for action consistent with the technological imperative—but that we also identify authority claims and suggest ways of contesting these claims. In short, we can "identify the relations of power that block" the desire for communicative action and "offer strategies for subverting that power, for betraying it into communicative action" [22, p. 264]. Wells' strategy for critical instruction can be supplemented with other strategies, such as Kate Ronald's proposal that students write *about* professional texts, examining discipline-specific constraints [8, p. 28]. Or the teacher may point out how the problem-solution pattern in technical reports implies a closed system; discuss the possibility of opening up a broader definition of criteria for writing proposals and feasibility reports; or suggest using a less impersonal style to bring the human element back into technical discourse.

However, it is important to go beyond teaching traditional structures from a critical perspective. If we claim the territory of political discourse as part of the province of technical discourse, we should teach students practical reasoning, that is, the process of deliberation and judgment that Garver describes in Teaching Writing and Teaching Virtue. Garver claims that practical reasoning goes beyond expressive and scientific writing, for the subject of practical reasoning is "contingent facts that can be other than they are, that action can do something about, that are worth worrying about" [17, p. 66]. While we probably already teach deliberation and judgment when we teach feasibility and investigation reports, we tend to do it within the constraints of an assumed audience—namely decision-makers within a company—and present private and governmental forums. Certainly, the power of audience over the writer is widely acknowledged, the most well known statement of this phenomenon being Perelman and Olbrechts-Tyteca's claim that the rhetor must always adapt the discourse to the audience [25]. Therefore, it is important that we open up the definition of audience to include the public; that is, we should incorporate at least some deliberative or judgmental discourse appropriate for a public forum.

But to do this, we need to create an imaginary society in which a public forum for such issues actually exists. It is at the point where we break with present reality, where we pretend that we live in an idealized society, that we begin to create a new social order. By writing for a public forum, even an imaginary one, students can begin to see the possible clash between the values of an audience in industry, heavily influenced by the profit motive, and the concerns of the public. Further, such writing works against the rhetoric of concealment by bringing issues before citizens and by calling into play value judgments that usually are not part of the decision-making process when deliberations about technology are confined to the privacy of an in-house report.

Redefining what we teach—that is, expanding the scope of technical communication to include public discourse about technology—would change programs as well as classes. Ph.D. programs in rhetoric and technical communication would begin to incorporate classes devoted to policy and to the philosophy of technology. This already happens in informal ways at places like Rensselaer Polytechnic Institute, where many students supplement their studies in rhetoric by taking classes from the Science and Technology Studies Department and by asking faculty from that department to sit on their committees. But a serious commitment to technical communication as a social act would eventually require that these sorts of studies be officially incorporated into the program, a direction presently being pursued in the new Ph.D. program in rhetoric and technical communication at Michigan Technological University.

## THE APPRENTICESHIP MODEL OF TEACHING

Not only do I suggest that we expand the definition of what we teach; I also suggest that we change the way we teach technical communication. Present practices often do not take seriously Miller's claim that teaching the course means enculturating students. As a result, we often teach the course as a skills course, creating a professional distance between ourselves and students, comparable to a seller-buyer relationship. After all, if all we are doing is teaching skills, we can impart what we know and never attempt to influence students. However, if we are enculturating students, if we are introducing them to the discourse community of industry and the larger discourse community of public citizenship, then the model offered by apprenticeship is more appropriate than the model offered by the market.

I am aware that some will object to the apprenticeship model. Marilyn Cooper's criticism of the concept of discourse communities applies to apprenticeship as well, for apprenticeship assumes that something like discourse communities exists and that the teacher initiates students into that social structure [26, p. 216]. Indeed, apprenticeship implies that the teacher represents the culture and that students learn through imitation [27, p. 53]. Reactions against this hierarchical system are understandable, especially since cultural systems have usually excluded or marginalized certain people. However, the alternative requires a commitment to expressive discourse, a form of discourse that excludes its practitioners simply because members of empowered communities perceive it as alien or unorthodox.

Therefore, we should make cautious use of apprenticeship as we employ it to bring students into the cultures that we represent. That is, even though we teach the discourse of the military-industrial complex, we can make clear that alternative cultures exist and that we identify with those cultures. Admittedly, such a view produces a rhetoric of conversion, but, after all, this is exactly what Ohmann calls for when he says that "we either teach politically . . . or we contribute to the mystification that so often in universities diverts and deadens the critical power" [1, p. 335]. The word of the teacher is somewhat alien to the world view of the students, but it is nevertheless an authoritative word; and as John Edlund points out in his analysis of Bakhtin, the teacher is a member of the social group that constitutes the class [28, p. 62]. Thus, we are in a position to help students appropriate and assimilate language practices about technology that go beyond the reductive structures of traditional technical genres.

There are many ways to apply the concept of apprenticeship to technical communication courses, but by way of example, I will briefly describe one system that I have been experimenting with. Since I identify primarily with the tradition of classical rhetoric (despite some of its social inequities, it nevertheless offers ways to subvert the technological mindset), I have adopted classical pedagogical practices that depend on imitation, a way of teaching I have discussed elsewhere [29]. I divide the course into two segments. During the first half of the course, I teach technical forms by asking students to do such things as copy, imitate, summarize, and transform examples of technical discourse. During class, we discuss these structures, and I link them to thought processes, pointing out how the structures exclude various considerations. This part of the class fairly closely resembles a traditional technical writing class, with two exceptions: students go through the forms rapidly by using the imitation exercises, and my discussion of the forms focuses primarily on their schematic nature and their function in social settings, rather than on details of correctness and usage.

The second segment of the course breaks with tradition. In an attempt to model the process of deliberation and judgment, I assign a single topic to the whole class, which is divided into two advocate groups and one arbitration group. Ideally, the topic is a question about a present policy decision, but because of the rhetoric of concealment that dominates our present discourse about technology, students find it difficult to get the information they need to build cases and decide issues. For instance, when I asked students to work on a local current issue—whether or not a paper mill should be built nearby—they soon ran out of information because the paper industry wanted to protect newly developed technology that they claimed could produce white paper without creating dioxins. Therefore, I choose a well-documented case from the past and ask students to take sides. My most successful assignment requires students to investigate the 1913 labor strikes in the copper-mining region where we live. The university's archives are rich in material on this subject, and students have access to information they would not be able to get about a current issue.

2

Part of the class is prolabor, part is procompany, and part is arbitration. Thus, the class as a whole models the deliberative process. During this time, I teach rhetorical concepts like *stasis* (how to determine the issue in a case), *kairos* (learning to take advantage of the opportune moment), and invention. I also teach students alternative genres for presenting their cases, such as the classical polemic speech and the Rogerian argument suggested by Richard Young, Alton Becker, and Kenneth Pike [30, p. 283]. Students go out on strike and participate in debates, and even the arbitration group writes majority and minority opinions. In this way, the total rhetorical exchange within the class functions to forge *prakton agathon*, "a concrete act of enlightened expediency," and the students engage in a modeled experience of performing a social act no longer constrained by present social restrictions [19, p. 23].

Defining technical communication as a practice has major significance for technical communication teachers. It allows us to see ourselves as doing more than teaching a set of skills, but it also places ethical and political responsibility upon us. If we continue to teach the course in traditional ways, we perpetuate a form of discourse that blocks social action; if we refuse to teach the conventions appropriate for industry, we fail to give our students the power they need to enter the dominant culture. Bizzell expresses the dilemma better than I can: "Our dilemma is that we want to empower students to succeed in the dominant culture so that they can transform it from within; but we fear that if they do succeed, their thinking will be changed in such a way that they will no longer want to transform it." However, by redefining the function and scope of technical communication, we may be able to teach it in such a way that students will be able to use technical genres and yet resist their power. We can even hope that a few among our students will find ways to transform present practices and open up opportunities for public social action.

## REFERENCES

1. R. Ohmann, *English in America: A Radical View of the Profession*, Oxford University Press, New York, 1976.
2. J. H. Mitchell, It's a Craft Course: Indoctrinate, Don't Educate, *The Technical Writing Teacher, 4*, pp. 2-6, 1976.
3. J. C. Mathes, D. W. Stevenson, and P. Klaver, Technical Writing: The Engineering Educator's Responsibility, *Engineering Education, 69*, pp. 331-334, 1979.
4. S. M. Halloran, Eloquence in a Technological Society, *Central States Speech Journal, 29*, pp. 221-227, 1978.
5. N. Johnson, Three Nineteenth-Century Rhetoricians: The Humanist Alternative to Rhetoric as Skills Management, in *The Rhetorical Tradition and Modern Writing*, J. J. Murphy (ed.), Modern Language Association, New York, 1982.

6. R. J. Connors, The Rise of Technical Writing Instruction in America, *Journal of Technical Writing and Communication, 12*, pp. 329-352, 1982.

7. C. R. Miller, A Humanistic Rationale for Technical Writing, *College English, 40*, pp. 610-617, 1979.

8. K. Ronald, The Politics of Teaching Professional Writing, *Journal of Advanced Composition, 7*, pp. 23-30, 1987.

9. J. Swales, *Approaching the Concept of Discourse Community*, Conference on College Composition and Communication, Atlanta, Georgia, March 1987.

10. C. R. Miller, Genre as Social Action, *Quarterly Journal of Speech, 70*, pp. 151-167, 1984.

11. J. Poulakos, Toward a Sophistic Definition of Rhetoric, *Philosophy and Rhetoric, 16*, pp. 35-48, 1983.

12. P. Bizzell, *What Is a "Discourse Community"?* Penn State Conference on Rhetoric and Composition, University Park, Pennsylvania, July 1987.

13. C. R. Miller, *What's Practical about Technical Writing*, in *Technical Writing: Theory and Practice*, B. E. Fearing and W. K. Sparrow (eds.), Modern Language Association, New York, 1989.

14. Aristotle, *Ethics*, J.A.K. Thompson (trans.), Penguin, New York, 1976.

15. B. Warnick, Judgment, Probability, and Aristotle's Rhetoric, *Quarterly Journal of Speech, 75*, pp. 299-311, 1989.

16. Aristotle, *Rhetoric*, W. Rhys Roberts (trans.), Modern Library, New York, 1954.

17. E. Garver, Teaching Writing and Teaching Virtue, *Journal of Business Communication, 22*, pp. 51-73, 1985.

18. L. S. Self, Rhetoric and *Phronesis*: The Aristotelian Ideal, *Philosophy and Rhetoric, 12*, pp. 1301-45, 1979.

19. O. L. Brownstein, Aristotle and the Rhetorical Process, in *Rhetoric: A Tradition in Transition*, W. R. Fisher (ed.), Michigan State University Press, Houghton, Michigan, 1974.

20. A. MacIntyre, *After Virtue: A Study in Moral Theory*, (2nd Edition), University of Notre Dame, Notre Dame, Indiana, 1984.

21. D. N. Dobrin, What's the Purpose of Technical Communication, *The Technical Writing Teacher, 12*, pp. 146-160, 1985.

22. S. Wells, Jurgen Habermas, Communicative Competence, and the Teaching of Technical Discourse, in *Theory in the Classroom*, C. Nelson (ed.), University of Illinois Press, Champaign, Illinois, 1986.

23. F. A. Rossini, Technology Assessment: A New Type of Science? *Research in Philosophy and Technology, 2*, pp. 341-355, 1979.

24. G. T. Goodnight, The Personal, Technical, and Public Spheres of Argument: A Speculative Inquiry into the Art of Public Deliberation, *Journal of the American Forensic Association, 18*, pp. 214-227, 1982.

25. C. Perelman and L. Olbrechts-Tyteca, *The New Rhetoric: A Treatise on Argumentation*, J. Wilkinson and P. Weaver (trans.), University of Notre Dame Press, Notre Dame, Indiana, 1969.

26. M. M. Cooper and M. Holzman, *Writing as Social Action*, Boynton, Portsmouth, New Hampshire, 1989.

27. M. Polanyi, *Personal Knowledge: Towards a Post-Critical Philosophy*, Harper, New York, 1964.
28. J. R. Edlund, Bakhtin and the Social Reality of Language Acquisition, *Writing Instructor, 7*, pp. 56-67, 1988.
29. D. L. Sullivan, Attitudes Toward Imitation: Classical Culture and the Modern Temper, *Rhetoric Review, 8*, pp. 5-21, 1989.
30. R. E. Young, A. L. Becker, and K. L. Pike, *Rhetoric: Discovery and Change*, Harcourt, New York, 1970.

# Contributors

JO ALLEN is associate professor in the Department of English at East Carolina University, Greenville, North Carolina. She is co-director of the Technical and Professional Communication program and teaches technical communication at the graduate level. Her principal current research interest is the role of social and political issues in shaping technical communication.

PAUL M. DOMBROWSKI is assistant professor in the Department of English at Ohio University in Athens, Ohio. He teaches technical writing, literature, and rhetorical theory. He has published several articles on ethics and social constructionism. He is currently working on the implications of postmodernism for technical communication, and on psychological approaches to rhetoric.

ALAN G. GROSS is professor in the Department of Rhetoric at the University of Minnesota-Twin Cities. He teaches rhetorical theory and criticism. He has published several articles on rhetoric of science and *The Rhetoric of Science*. He is currently working on books on the rhetorical history of the scientific article and on Aristotle's rhetoric.

R. ALLEN HARRIS is associate professor in the Department of English at the University of Waterloo, Waterloo, Ontario. He teaches rhetoric, technical communications, and linguistics. He has published *The Linguistics Wars* on rhetoric within the field of linguistics and has a book in progress on landmark essays in rhetoric of science.

MARY M. LAY is professor of scientific and technical communication in the Department of Rhetoric at the University of Minnesota-Twin Cities. She is past president of the Association of Teachers of Technical Writing and current co-editor of the ATTW journal *Technical Communication Quarterly*. Several of her publications on gender issues and on collaborative writing, including *Collaborative Writing in Industry: Investigations in Theory and Practice,* have won awards.

MICHAEL H. MARKEL is associate professor in the Department of English at Boise State University in Boise, Idaho. He is Director of Technical Communication in an expanding program and teaches several undergraduate and graduate courses in technical communication. He has published several articles on ethics and on social constructionism, as well as *Writing in the Technical Fields* and a textbook, *Technical Writing: Situations and Strategies.*

CAROLYN R. MILLER is professor in the Department of English at North Carolina State University in Raleigh, North Carolina. Her article, *A Humanistic Rationale for Technical Writing*, is probably the most frequently cited article in technical writing. She is working on a book about the rhetoric of technology and is president-elect of the Rhetoric Society of America.

DALE L. SULLIVAN is assistant professor in the Department of English at Northern Illinois University at DeKalb Illinois. He is a generalist in rhetorical theory and also interested in applications in technical writing, the rhetoric of science, and writing across the curriculum.

# Index

Acker, J., 133
Allen, J., 125, 136-138
  Gender Issues in Technical
  Communication Studies: An
  Overview of the Implications for
  the Profession, Research, and
  Pedagogy, 161-169
Aristotle, 6
Arts, Liberal, 3

Barton, B. and M. Barton, 193
Bazerman, C., 9, 18, 85, 87, 92-93
Berger, P. L. and T. Luckmann, 83
Bizzell, P., 184
Bleier, R., 127, 128, 129
Blyler, N. R. and C. Thralls, 86
Bradford, D., 9
Brockmann, R. J. and Fern Rook,
  188, 190
Bronowski, J., 6
Bruffee, K., 84, 85
Buchholz, W. J., 186, 190-191
Buker, E. A., 131
Burke, K., 184

Cain, B. E., 7
Cancian, F. M., 133
Christians, C., 189
Chodorow, N., 126

Clark, G., 187, 188
Cozzens, S. E. and T. F. Gieryn, 84

Dewey, J., 9, 132
Dilthey, W., 24
Dobrin, D., 183
Doheny-Farina, S., 188
Dombrowski, P. M., 10-11, 86-87,
  91-92, 187, 192
  Challenger and the Social Contin-
  gency of Meaning: Two Lessons
  for the Technical Communication
  Classroom, 97-109
Dragga, S., 134
Dreyfus, H. L. and J. Rouse, 24
Dualism, Traditional, 5-8
Durbin, P. T., 191

Einstein, A., 88
Ellul, J., 9, 182, 192
Ethics, 181-198
Evernden, N., 91

Faigley, L., 85
Fee, E., 129
Feminist Critiques of Science and
  Gender Issues, 125-139
Feyerabend, P., 10, 18, 83, 84
Fleck, L., 10

Geertz, C., 83
Gergen, K. J., 83
Gilligan, C., 130, 183
Gorelick, S., 132
Gross, A. G., 4, 17, 19, 28-29, 183
   Discourse on Method: The
   Rhetorical Analysis of Scientific
   Texts, 63-79
Guigon, C. B., 17

Hagendijk, R., 84
Halloran, S. M., 4, 7, 9, 19, 20, 28,
   83, 184
Halloran, S. M. and M. Whitburn,
   18, 83
Harding, S., 130
Harris, E., 9
Harris, R. A., 15, 23, 27
   Assent, Dissent, and Rhetoric in
   Science, 33-62
Heldke, L., 132
Herzberg, B., 22
Hiley, D. R., J. F. Bohman and
   R. Shusterman, 24, 91
Hirsch, M. and E. F. Keller, 128,
   129, 133
Holism, Contemporary, 8-11
Humanism, 2-5

Isocrates, 6

Jaeger, W., 3
Jensen, J. V., 189
Johannesen, R. L., 188
Johnson, N., 192, 195
Journet, D., 21, 85

Katz, S., 187, 192
Keller, E. F., 89, 131, 132
Kennedy, G. A., 3
Kent, T., 85, 89
Kerferd, G. B., 81

Kimball, B., 3, 6, 7, 9
Knorr-Cetina, K., 83
Kohanski, A., 6
Kremers, M., 191-192
Kuhn, T., 10, 82, 88

Larrabbe, M. J., 183
Latour, B., 26
Latour, B. and S. Woolgar, 83
Lay, M. M., 125, 126, 131, 134-136
   Feminist Theory and the Redefini-
   tion of Technical Communica-
   tion, 141-159
Lay, M. M. and W. M. Karis, 85
LeFevre, K. B., 9, 85
Levinas, E., 193
Longino, H. E., 131
Longino, H. E. and E. Hammonds, 127
Lorber, J. and S. A. Farrell, 81, 129

Mackin, J. A., Jr., 22
Markel, M., 186, 194
   A Basic Unit on Ethics for Tech-
   nical Communication, 199-221
Mead, G. H., 83
Merton, R. K., 83
Miller, C. R., 1, 9, 18, 22, 82, 83, 85,
   90, 92-93, 184
   Some Perspectives on Rhetoric,
   Science, and History, 111-123
Miller, C. R. and J. Selzer, 86
Monsma, S., 182
Moore, P., 189

Namenwirth, M., 129
Nelson, J. S. and Megill, A., 3, 25
Noddings, N., 183
Nussbaum, M., 81

Odell, L., 85
Odell, L. and D. Goswami, 85
Ornatowski, C. M., 189

Paradis, J., D. Dobrin, and R. Miller, 86
Parsons, G., 9
Pickering, A., 82
Pitts, J. C., 191
Plato, 6
Positivism, 83
Possin, K., 192
Prelli, L. J., 23
Protagoras, 3-4, 93

Quine, W. V. and J. S. Ullian, 83

Rhetoric of Science, 15-59
Rorty, R., 7, 83, 132
Rose, H., 129
Rouse, J., 24
Rubens, P., 2
Rude, C. D., 26
Rutter, R., 8, 9, 21

Sachs, H., 191
Samuels, M. S., 6, 9
Sanger, D. E., 187
Sauer, B. A., 134
Sawyer, T. M., 189-190
Simons, H. W., 23-24, 81
Singer, P., 181
Snow, C. P., 8, 9
Social Constructionism, 81-96
Sturges, D. L., 189

Sullivan, D., 9, 194-195
    Political-Ethical Implications of
    Defining Technical Communica-
    tion as a Practice, 223-234

Thralls, C. and N. R. Blyler, 86
Trainer, T., 183
Tuana, N., 128, 130

United States, Presidential Com-
    mission on the Space Shuttle
    Challenger Accident ("Rogers
    Commission"), 90, 185-186
United States, Congress, House,
    Investigations of the Challenger
    Accident, 90

von Glaserfeld, E., 81, 88

Waddell, C., 20, 84
Weaver, R., 15, 182, 184
Weimer, W. B., 10, 28
Whitburn, M., 9, 184
Winner, L., 9
Winsor, D., 86, 187
Wright, F. L., 185
Wylie, A., K. Okruhlik, L. Thielen-
    Wilson, and S. Morton, 128, 133

Zappen, J. P., 21, 191
Zucker, A., 181